Student Mathematics Handbook

KARL J. SMITH

Calculus
SECOND EDITION

Gerald L. Bradley Karl J. Smith

PRENTICE HALL, Upper Saddle River, NJ 07458

Executive Editor: George Lobell
Special Projects Manager: Barbara A. Murray
Production Editor: Shea Oakley
Supplement Cover Manager: Paul Gourhan
Supplement Cover Designer: Liz Nemeth
Manufacturing Buyer: Alan Fischer
Editorial Assistant: Gale A. Epps

 © 1998 by Prentice-Hall, Inc.
Simon & Schuster / A Viacom Company
Upper Saddle River, NJ 07458

Printed in the United States of America

10 9 8 7 6 5 4 3 2

ISBN 0-13-081954-9

Prentice-Hall International (UK) Limited, *London*
Prentice-Hall of Australia Pty. Limited, *Sydney*
Prentice-Hall Canada, Inc., *London*
Prentice-Hall Hispanoamericana, S.A., *Mexico*
Prentice-Hall of India Private Limited, *New Delhi*
Prentice-Hall of Japan, Inc., *Tokyo*
Simon & Schuster Asia Pte. Ltd., *Singapore*
Editora Prentice-Hall do Brazil, Ltda., *Rio de Janeiro*

CONTENTS

PREFACE

Calculus is probably the first mathematics course you have taken that is not self-contained in the sense that the material from previous mathematics courses is expected *without* specifically mentioning it in the textbook. This supplement is a convenient reference book to be used along with your textbook, *Calculus,* to remind you of those formulas or topics that you may have forgotten.

This handbook is organized so that it can be used in two ways. The first use is as a reference manual providing a summary of terminology, formulas, and tables, not only of prerequisite mathematics, but also of the material covered in a standard calculus course. The second use is as a brief review of material assumed as a prerequisite for a course in calculus. This material is presented with examples, brief written exposition, and practice problems. The topics that are included with exposition and practice problems are sometimes missing from the backgrounds of many students who otherwise have the prerequisites for calculus. We remind the student that nearly everyone qualified to enroll in calculus has, for a variety of reasons, gaps in knowledge of prerequisite material, and the brief review in this handbook can help to bridge that gap. We provide this supplement to *Calculus* free of charge with the purchase of the textbook, in an attempt to help ensure your success in calculus. Use this book for reference, and as a handbook as you progress through the course.

In addition, new technology has changed the emphasis of many of the topics in a calculus course. One recent change is the acknowledgement of calculators and computers to help not only with the mechanics of algebra, but also with the mechanics of differentiation and integration in calculus. Outside of the academic environment, engineers and physicists tells us that using available technology, as well as tables of integration, is by far more important than many of the esoteric topics they were taught in their calculus courses many years ago. For that reason, calculus books are evolving, and the emphasis is not on obscure esoteric topics, but rather practical knowledge that balances between application and theory. To capture this new emphasis, we see the need for you to have a complete integration table, so one has been provided in this handbook.

CHAPTER 1

Review of Geometry

In this book we use the following variables when stating formulas: A = area, P = perimeter, C = circumference, S = surface area, and V = volume. Also, r denotes radius, h altitude, l slant height, b base, B area of base, θ central angle expressed in radians.

1.1

Polygons

CLASSIFICATION

Type	Number of sides
triangle	3
quadrilateral	4
pentagon	5
hexagon	6
heptagon	7
octagon	8
nonagon	9
decagon	10
undecagon	11
dodecagon	12

TRIANGLES

$A = \frac{1}{2}bh$

$P = a + b + c$

The sum of the measures of the angles of a triangle is 180°.

Pythagorean theorem   If angle C is a right angle, then $c^2 = a^2 + b^2$.

45°-45°-90° triangle theorem   For any right triangle with acute angles measuring 45°, the legs are the same length, and the hypotenuse has a length equal to $\sqrt{2}$ times the length of one of those legs.

30°-60°-90° triangle theorem   For any right triangle with acute angles measuring 30° and 60°:
1. The hypotenuse is twice as long as the leg opposite the 30° angle (the shorter leg).
2. The leg opposite the 30° angle (the shorter leg) is $\frac{1}{2}$ as long as the hypotenuse.
3. The leg opposite the 60° angle (the longer leg) equals the length of the other (shorter) leg times $\sqrt{3}$.
4. The leg opposite the 30° angle equals the length of the other leg divided by $\sqrt{3}$.

Equilateral triangle   For any equilateral triangle:

$\alpha = \beta = \gamma = 60°$ $A = \frac{1}{4}b^2\sqrt{3}$ $h = \frac{1}{2}b\sqrt{3}$

QUADRILATERALS

Rectangle	**Square**	**Parallelogram**	**Trapezoid**
$A = \ell w$	$A = s^2$	$A = bh = ab\sin\theta$	$A = \frac{1}{2}h(a + b)$
$P = 2\ell + 2w$	$P = 4s$	$P = 2a + 2b$	$P = a + b$
			$\quad + h(\csc\theta + \csc\phi)$

Diagonal $= \sqrt{\ell^2 + w^2}$ Diagonal $= s\sqrt{2}$

REGULAR POLYGON OF n SIDES
$$A = \tfrac{1}{4}na^2\cot\tfrac{\pi}{n}$$

$$P = an$$

central angle: $\dfrac{2\pi}{n}$

1.2

Circles

TERMINOLOGY

Definition: In a plane, a **circle** is the set of all points a given distance, called the **radius**, from a given point, called the **center**.

Circumference: distance around a circle.

Chord: a line joining two points of a circle.

Diameter: a chord through the center: AB is Figure 1.1.

Arc: part of a circle: BC, AC, or ACB in Figure 1.1. The length s of an arc of a circle of radius r with central angle θ (measured in radians) is $s = r\theta$.

To *intercept an arc* is to cut off the arc; in Figure 1.1, $\angle COB$ intercepts BC.

Tangent of a circle is a line that intersects the circle at one and only one point. **Figure 1.1**

Secant of a circle is a line that intersects the circle at exactly two points.

Inscribed polygon is a polygon all of whose sides are chords of a circle. A regular inscribed polygon is a polygon all of whose sides are the same length.

Inscribed circle is a circle to which all the sides of a polygon are tangents.

Circumscribed polygon is a polygon all of whose sides are tangents to a circle.

Circumscribed circle is a circle passing through each vertex of a polygon.

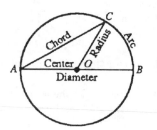

BASIC FORMULAS

Circle	Sector	Segment

Circle

$A = \pi r^2$
$C = 2\pi r = \pi d$

Sector

$A = \frac{1}{2}r^2\theta$
$s = r\theta$

Segment

$A = \frac{1}{2}r^2(\theta - \sin\theta)$

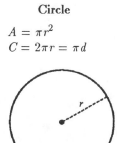

1.3

Solid Geometry

Rectangular parallelepiped (box)

$V = abc$

Diagonal $= \sqrt{a^2 + b^2 + c^2}$.

Prism

$V = Bh$

B is area of the base

Pyramid

$V = \frac{1}{3}Bh$

B is area of the base

Hexagonal prism

Tetrahedron

(a pyramid with a triangular base)

$V = \frac{1}{3}h\sqrt{s(s-a)(s-b)(s-c)}$

where $s = \frac{1}{2}(a + b + c)$

Right circular cylinder

$V = \pi r^2 h$
Lateral surface $= 2\pi rh$

$S = 2\pi rh + 2\pi r^2$

Right circular cone

$V = \frac{1}{3}\pi r^2 h$
Lateral surface $= \pi rl$

$S = \pi rl + \pi r^2$

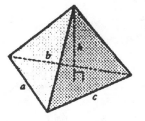

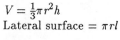

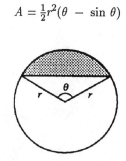

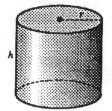

Frustum of a right circular cone

$V = \frac{1}{3}\pi h(r^2 + rR + R^2)$ or

$V = \frac{1}{3}h(B_1 + \sqrt{B_1 B_2} + B_2)$

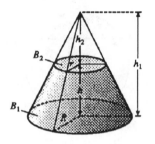

Frustum of a pyramid

$V = \frac{1}{3}h(B_1 + \sqrt{B_1 B_2} + B_2)$

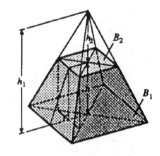

Torus

$S = 4\pi^2 Rr$

$V = 2\pi^2 Rr^2$

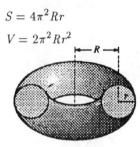

Spherical segment

$S = 2\pi rh$, with radius r

$V = \frac{1}{6}\pi h(3r_1{}^2 + 3r_2{}^2 + h^2)$

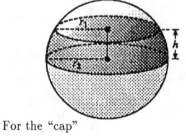

For the "cap"
$S = 2\pi rh$
$V = \frac{\pi}{6}h(3r_2{}^2 + h)$

Cylinder with a cross-sectional area A
$V = Ah; \quad S = p\ell + 2A$

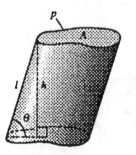

Prismatoid, pontoon, wedge
$V = \frac{1}{6}h(B_0 + 4B_1 + B_2)$

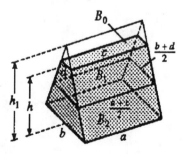

Quadric Surfaces

Sphere

$$x^2 + y^2 + z^2 = r^2$$

$$V = \tfrac{4}{3}\pi r^3$$
$$S = 4\pi r^2$$

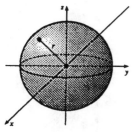

Ellipsoid

$$\frac{x^2}{a^2} + \frac{y^2}{b^2} + \frac{z^2}{c^2} = 1$$

$$V = \tfrac{4}{3}\pi abc$$

Elliptic Paraboloid

$$\frac{x^2}{a^2} + \frac{y^2}{b^2} = z$$

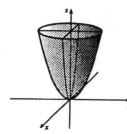

Hyperboloid of One Sheet

$$\frac{x^2}{a^2} + \frac{y^2}{b^2} - \frac{z^2}{c^2} = 1$$

Hyperboloid of Two Sheets

$$\frac{x^2}{a^2} - \frac{y^2}{b^2} - \frac{z^2}{c^2} = -1$$

Hyperbolic paraboloid

$$\frac{y^2}{a^2} - \frac{x^2}{b^2} = z$$

Oblate spheroid is formed by the rotation of the ellipse $\frac{x^2}{a^2} + \frac{y^2}{b^2} = 1$ about its minor axis, b. Let ϵ be the eccentricity. $\qquad V = \tfrac{4}{3}\pi a^2 b \qquad\qquad S = 2\pi a^2 + \pi\frac{b^2}{\epsilon}\ln\!\left(\frac{1+\epsilon}{1-\epsilon}\right)$

Prolate spheroid is formed by the rotation of the ellipse $\frac{x^2}{a^2} + \frac{y^2}{b^2} = 1$ about its minor axis, a. Let ϵ be the eccentricity. $\qquad V = \tfrac{4}{3}\pi ab^2 \qquad\qquad S = 2\pi b^2 + \pi\frac{ab}{\epsilon}\sin^{-1}\epsilon$

1.4

Congruent Triangles

We say that two figures are **congruent** if they have the same size and shape. For **congruent triangles** ABC and DEF, denoted by $\triangle ABC \simeq \triangle DEF$, we may conclude that all six corresponding parts (three angles and three sides) are congruent. We call these **corresponding parts.**

EXAMPLE 1

Corresponding parts of a triangle

Name the corresponding parts of the given triangles.
a. $\triangle ABC \simeq \triangle A'B'C'$ **b.** $\triangle RST \simeq \triangle UST$

Solution

a.
$\overline{AB}$ corresponds to $\overline{A'B'}$
$\overline{AC}$ corresponds to $\overline{A'C'}$
$\overline{BC}$ corresponds to $\overline{B'C'}$
$\angle A$ corresponds to $\angle A'$
$\angle B$ corresponds to $\angle B'$
$\angle C$ corresponds to $\angle C'$

b.
$\overline{RS}$ corresponds to $\overline{US}$
$\overline{RT}$ corresponds to $\overline{UT}$
$\overline{ST}$ corresponds to $\overline{ST}$
$\angle R$ corresponds to $\angle U$
$\angle RTS$ corresponds to $\angle UTS$
$\angle RST$ corresponds to $\angle UST$

□

Line segments, angles, triangles, or other geometric figures are *congruent* if they have the same size and shape. In this section, we focus on triangles.

**CONGRUENT
TRIANGLES**

Two triangles are **congruent** if their corresponding sides have the same length and their corresponding angles have the same measure.

In order to prove that two triangles are congruent, you must show that they have the same size and shape. It is not necessary to show that all six parts (three sides and three angles) are congruent; if some of these six parts are congruent, it necessarily follows that the other parts are congruent. Three important properties are used to show triangle congruence:

**CONGRUENT
TRIANGLE
PROPERTIES**

SIDE–SIDE–SIDE (SSS)
If three sides of one triangle are congruent to three sides of another triangle, then the two triangles are congruent.

SIDE–ANGLE–SIDE (SAS)
If two sides of one triangle and the angle between those sides are congruent to the corresponding sides and angle of another triangle, then the two triangles are congruent.

ANGLE–SIDE–ANGLE (ASA)
If two angles and the side that connects them on one triangle are congruent to the corresponding angles and side of another triangle, then the two triangles are congruent.

EXAMPLE 2   **Finding congruent triangles**

Determine if each pair of triangles is congruent. If so, cite one of the congruent triangle properties.

Solution

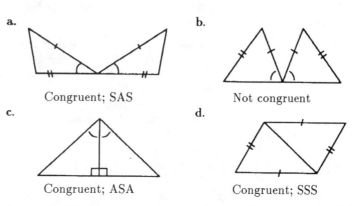

a.

Congruent; SAS

b.

Not congruent

c.

Congruent; ASA

d.

Congruent; SSS

A side that is in common to two triangles obviously is equal in length to itself, and it does not need to be marked.   ☐

In geometry, the main use of congruent triangles is when we want to know whether an angle from one triangle is congruent to an angle from a different triangle, or when we want to know whether a side from one triangle is the same length as the side from another triangle. In order to do this, we often prove that one triangle is congruent to the other (using one of the three congruent triangle properties) and then use the following property:

**CONGRUENT TRIANGLE
PROPERTY**

Corresponding parts of congruent triangles are congruent.

1.5

Similar Triangles

It is possible for two figures to have the same shape, but not necessarily the same size. These figures are called **similar** figures. We will now focus on **similar triangles.** If $\triangle ABC$ is similar to $\triangle DEF$, we write

$$\triangle ABC \sim \triangle DEF$$

Similar triangles are shown in Figure 1.2.

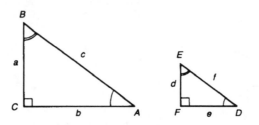

Figure 1.2 Similar triangles

You should note that congruent triangles must be similar, but similar triangles are not necessarily congruent. Since similar figures have the same shape, we talk about **corresponding angles** and **corresponding sides**. The corresponding angles of similar triangles are the angles that have the same measure. It is customary to label the vertices of triangles with capital letters and the sides opposite the angles at those vertices with corresponding lower case letters. It is easy to see that, if the triangles are similar, the corresponding sides are the sides opposite equal angles. In Figure 1.2, we see that

$\angle A$ and $\angle D$ are corresponding angles;
$\angle B$ and $\angle E$ are corresponding angles; and
$\angle C$ and $\angle F$ are corresponding angles.

Side a $(\overline{BC})$ is opposite $\angle A$, and d $(\overline{EF})$ is opposite $\angle D$, so we say
    that a corresponds to d;
b corresponds to e; and
c corresponds to f.

Even though corresponding angles are the same size, corresponding sides do not need to have the same length. If they do have the same length, then the triangles are congruent. However, when they are not the same length, we can say they are *proportional*. As Figure 1.2 illustrates, when we say the sides are proportional, we mean

$$\frac{a}{b} = \frac{d}{e} \qquad\qquad \frac{a}{c} = \frac{d}{f} \qquad\qquad \frac{b}{c} = \frac{e}{f}$$

$$\frac{b}{a} = \frac{e}{d} \qquad\qquad \frac{c}{a} = \frac{f}{d} \qquad\qquad \frac{c}{b} = \frac{f}{e}$$

SIMILAR TRIANGLES

> Two triangles are **similar** if two angles of one triangle have the same measure as two angles of the other triangle. If the triangles are similar, then their corresponding sides are proportional.

EXAMPLE 1

Similar triangles

Identify pairs of triangles that are similar in Figure 1.3.

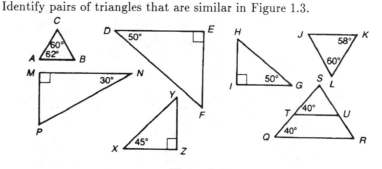

Figure 1.3

Solution $\triangle ABC \sim \triangle JKL; \quad \triangle DEF \sim \triangle GIH; \quad \triangle SQR \sim \triangle STU.$ □

EXAMPLE 2

Finding unknown lengths in similar triangles

Given the similar triangles below, find the unknown lengths marked b' and c'.

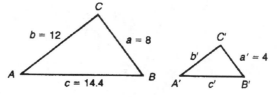

Solution   Since corresponding sides are proportional (other proportions are possible), we have

$$\frac{a'}{a} = \frac{b'}{b} \qquad\qquad\qquad \frac{a}{c} = \frac{a'}{c'}$$

$$\frac{4}{8} = \frac{b'}{12} \qquad\qquad\qquad \frac{8}{14.4} = \frac{4}{c'}$$

$$b' = \frac{4(12)}{8} \qquad\qquad\qquad c' = \frac{14.4(4)}{8}$$

$$= 6 \qquad\qquad\qquad\qquad = 7.2 \qquad\qquad\qquad □$$

EXAMPLE 3

Finding a perimeter using similar triangles

In equilateral $\triangle ABC$, suppose $DE = 2$ and is parallel to $\overline{AB}$ as shown at right. If $\overline{AB}$ is three times as long as DE, what is the perimeter of quadrilateral $ABED$?

Solution

$\triangle ABC \sim \triangle DEC$ so $\triangle DEC$ is equilateral. This means that $\overline{CE}$ and $\overline{DC}$ both have length 2; thus, $\overline{EB}$ and $\overline{AD}$ both have length 4. The perimeter of the quadrilateral is

$$|\overline{AB}| + |\overline{BE}| + |\overline{DE}| + |\overline{AD}| = 6 + 4 + 2 + 4 = 16 \qquad \square$$

Finding similar triangles is simplified even further if we know the triangles are right triangles, because then the triangles are similar if one of the acute angles has the same measure as an acute angle of the other.

EXAMPLE 4

Using similar triangles to find an unknown length

Suppose that a tree and a yardstick are casting shadows as shown in Figure 1.4. If the shadow of the yardstick is 3 yards long and the shadow of the tree is 12 yards long, use similar triangles to estimate the height of the tree if you know that angles S and S' are the same size.

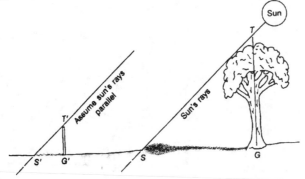

Figure 1.4

Solution

Since $\angle G$ and $\angle G'$ are right angles, and since S and S' are the same size, we see that $\triangle SGT \sim \triangle S'G'T'$. Therefore, corresponding sides are proportional.

$$\frac{1}{3} = \frac{h}{12}$$
$$h = \frac{1(12)}{3}$$
$$= 4$$

The tree is 4 yards tall. $\qquad \square$

1.6 PROBLEM SET 1

A

1. In $\triangle TRI$ and $\triangle ANG$ shown at
    right, $\angle R \simeq \angle N$ and $|\overline{TR}| = |\overline{AN}|$.
    Name other pairs you would need
    to know in order to show that the
    triangles are congruent by
    a. SSS   **b.** SAS   **c.** ASA

2. In $\triangle ABC$ and $\triangle DEF$ shown at right,
    $\angle A \simeq \angle D$ and $|\overline{AC}| = |\overline{DF}|$.
    Name other pairs you would need to
    know in order to show that the
    triangles are congruent by
    a. SSS   **b.** SAS   **c.** ASA

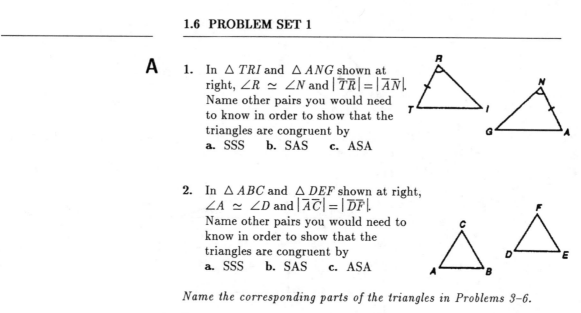

Name the corresponding parts of the triangles in Problems 3–6.

3.

4.

5.

6.

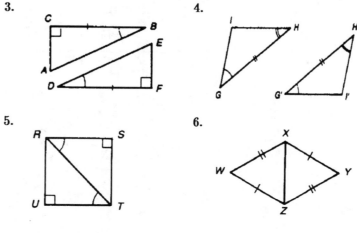

*In Problems 7–10, determine if each pair of triangles is congruent.
If so, cite one of the congruent triangle properties.*

7.

8.

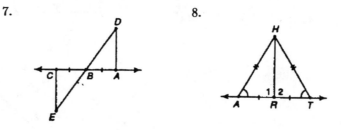

9.

10.

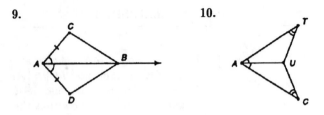

In Problems 11–16, tell whether it is possible to conclude that the pairs of triangles are similar.

11.

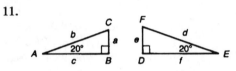

12.

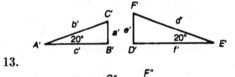

13.

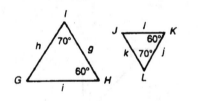

14.

15.

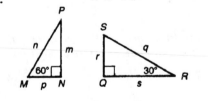

16.

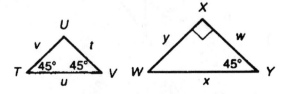

B   *Given the two similar triangles shown, find the unknown lengths in Problems 17–22.*

17. $a = 4$, $b = 8$, $a' = 2$; find b'.   **18.** $b = 5$, $c = 15$, $b' = 2$; find c'.
19. $c = 6$, $a = 4$, $c' = 8$; find a'.   **20.** $a' = 7$, $b' = 8$, $a = 5$; find b.
21. $b' = 8$, $c' = 12$, $c = 4$; find b.   **22.** $c' = 9$, $a' = 2$, $c = 5$; find a.
23. How far from the base of a building must a 26-ft ladder be placed so that it reaches 10 ft up the wall?
24. How high up a wall does a 26-ft ladder reach if the bottom of the ladder is placed 6 ft from the building?
25. A carpenter wants to be sure that the corner of a building is square and measures 6 ft and 8 ft along the sides. How long should the diagonal be?
26. What is the exact length of the hypotenuse if the legs of a right triangle are 2 in. each?
27. What is the exact length of the hypotenuse if the legs of a right triangle are 3 ft each?
28. An empty lot is 40 ft by 65 ft. How many feet would you save by walking diagonally across the lot instead of walking the length and width?
29. A television antenna is to be erected and held by guy wires. If the guy wires are 15 ft from the base of the antenna and the antenna is 10 ft high, what is the exact length of each guy wire? What is the length of each guy wire rounded to the nearest foot? If three guy wires are attached, how many feet of wire should be purchased, if it cannot be bought in fractions of a foot?
30. In equilateral $\triangle ABC$, shown in the margin, D is the midpoint of segment $\overline{AB}$. What is the length of $\overline{CD}$?
31. In the figure shown at the right, AB and DE are parallel. What is the length of AB?
32. Ben walked diagonally across a rectangular field that measures 100 ft by 240 ft. How far did Ben walk?
33. Use similar triangles and a proportion to find the length of the lake shown in the figure below.

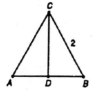

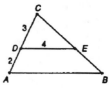

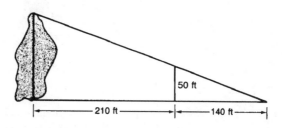

34. Use similar triangles and a proportion to find the height of the house shown in the figure below.

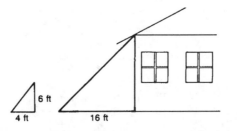

35. If a tree casts a shadow of 12 ft at the same time a 6-ft person casts a shadow of $2\frac{1}{2}$ ft, find the height of the tree to the nearest foot.

36. If a circular cone of height 10 cm and a radius of 4 cm contains liquid with height measuring 3.8 cm, what is the volume of the liquid?

37. A person 6 ft tall is walking away from a streetlight 20 ft high at the rate of 7 ft/s. How long (to the nearest ft) is the person's shadow at the instant when the person is 10 ft from the base of the lamppost? (See Example 2 of Section 3.7 of the text.)

38. A bag is tied to the top of a 5-m ladder resting against a vertical wall. Suppose the ladder begins sliding down the wall in such a way that the foot of the ladder is moving away from the wall. How high is the bag at the instant when the base of the ladder is 4 m from the base of the wall? (See Example 3 of Section 3.7 of the text.)

39. A tank filled with water is in the shape of an inverted cone 20 ft high with a circular base (on top) whose radius is 5 ft. How much water does the tank hold when the water level is 8 ft deep? (See Example 5 of Section 3.7 of the text.)

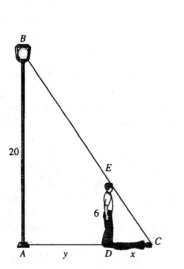

CHAPTER 2

Review of Algebra

2.1

Real Numbers

There are various sets of numbers with which you should be familiar.

Name	Symbol	Set	Examples
counting numbers or natural numbers	$\mathbb{N}$	$\{1, 2, 3, 4, \cdots\}$	72; $2{,}345{,}950$; $\sqrt{25}$; $\sqrt{1}$
Whole numbers	$\mathbb{W}$	$\{0, 1, 2, 3, \cdots\}$	0; 72; $\frac{10}{2}$; $\frac{16{,}425}{25}$
Integers	$\mathbb{Z}$	$\{\cdots, -2, -1, 0, 1, 2, \cdots\}$	-5; 0; $\frac{-10}{2}$; $9{,}562$; -82
Rational numbers	$\mathbb{Q}$	Numbers that can be written in the form $\frac{p}{q}$ where p and q are integers with $q \neq 0$	$\frac{1}{3}$; $\frac{4}{17}$; $.863214$; $.8666\cdots$; 5; 0; $\frac{-11}{3}$; -15; $\sqrt{\frac{1}{4}}$; 3.1416; $8.\overline{6}$
Irrational numbers	$\mathbb{Q}'$	Numbers whose decimal representations do not terminate and do not repeat	$5.123456789101112\cdots$; $\sqrt{2}$; $\sqrt{3}$; $4.31331333133331\cdots$; π; $\frac{\pi}{2}$
Real numbers	$\mathbb{R}$	The collection of all rational and irrational numbers	All examples listed above are real numbers. Not all numbers are real numbers, however. There is another set of numbers (not considered in this handbook) called *complex numbers*, which includes the real numbers as a subset.

2.2

Powers and roots

EXPONENTS

Definition of Exponent

If n is a positive integer, then $x^n = \underbrace{x \cdot x \cdot x \cdots x}_{n \text{ factors}}$, and $x^0 = 1$; $x^{-n} = \frac{1}{x^n}$

If m and n are integers, then

$$x^{\frac{1}{n}} = \sqrt[n]{x} \quad \text{whenever } \sqrt[n]{x} \text{ is defined}$$

and

$$x^{\frac{m}{n}} = (x^{\frac{1}{n}})^m = (\sqrt[n]{x})^m \quad \text{whenever } \sqrt[n]{x} \text{ is defined}$$

Factorial numbers

$$n! = n(n-1)(n-2) \cdots 3 \cdot 2 \cdot 1 \qquad\qquad 0! = 1 \qquad \text{for } n \text{ a nonnegative integer}$$

ROOTS

Recall that for any positive *even* integer n (called the *index* of the radical) and any positive number x (called the *radicand*),

$$y = \sqrt[n]{x} \text{ if and only if } y > 0 \text{ and } y^n = x$$

We call y *the positive nth root of x*. For example, the positive fourth root of 16 is denoted by $\sqrt[4]{16}$; we write $\sqrt[4]{16} = 2$ since $2^4 = 16$.

For any positive *odd* integer n any number x (positive or negative),

$$y = \sqrt[n]{x} \text{ and only if } y^n = x$$

and y is called the *nth root* of x. For example, $\sqrt[3]{-8} = -2$ since $(-2)^3 = -8$.

Note that $\sqrt{x^2} = |x|$ for any number x.

LAWS OF EXPONENTS

If r and s are real numbers, then

$$x^r \cdot x^s = x^{r+s}$$

$$(x^r)^s = x^{rs} \quad \text{whenever } x^r \text{ is meaningful}$$

$$(xy)^r = x^r y^r \quad \text{whenever } x^r \text{ and } y^r \text{ are meaningful}$$

$$\left(\frac{x}{y}\right)^r = \frac{x^r}{y^r} \quad \text{whenever } x^r \text{ and } y^r \text{ are meaningful and } y^r \neq 0$$

$$\frac{x^r}{x^s} = x^{r-s} \quad \text{whenever } x^r \text{ and } x^s \text{ are meaningful and } x^s \neq 0$$

FACTORS AND EXPANSIONS

Difference of squares: $\qquad a^2 - b^2 = (a - b)(a + b)$

Perfect square: $\qquad (a + b)^2 = a^2 + 2ab + b^2$

Difference of cubes: $\qquad a^3 - b^3 = (a - b)(a^2 + ab + b^2)$

Sum of cubes: $\qquad a^3 + b^3 = (a + b)(a^2 - ab + b^2)$

Perfect cube: $\qquad (a + b)^3 = a^3 + 3a^2b + 3ab^2 + b^3$

Binomial theorem: $\qquad (a + b)^n = a^n + \binom{n}{1}a^{n-1}b + \binom{n}{2}a^{n-2}b^2 + \cdots \quad$ where

$$\binom{n}{1} = \frac{n}{1}, \binom{n}{2} = \frac{n(n-1)}{1 \cdot 2}, \cdots, \binom{n}{r} = \frac{n!}{r!(n-r)!}, \cdots.$$

RADICALS

An algebraic expression containing radicals is **simplified** if all four of the following conditions are satisfied.

1. When the radicand is written in completely factored form, there is no factor raised to a power greater than or equal to the index of the radical.
2. No radical appears in a denominator.
3. No fraction (or negative exponent) appears within a radical.
4. There is no common factor (other than 1) between the exponent of the radicand and the index of the radical.

LOGARITHMS

Properties of logarithms

$\log (MN) = \log M + \log N$

$\log (M/N) = \log M - \log N$

$\log M^n = n \log M$

$\log \sqrt[n]{M} = \frac{1}{n} \log M$

$\log_b b = 1$

$\log_n 1 = 0.$

Special bases:

$\log x = \log_{10}x$; this is called a common logarithm

$\ln x = \log_e x$ where $e \approx 2.71828182845905$; this

$\qquad$ is called a natural logarithm

EXAMPLE 1	**Using laws of exponents**

$$\left(2x^2y^3\right)^4 = 2^4 x^{2 \cdot 4} \, y^{3 \cdot 4} = 16x^8y^{12} \qquad \qquad \square$$

EXAMPLE 2   **Simplifying radical expressions**

$$\sqrt[3]{\frac{x^6 y^7}{z^9}} = \left(\frac{x^6 y^7}{z^9}\right)^{1/3} = \frac{x^{6(1/3)} y^{7(1/3)}}{z^{9(1/3)}} = \frac{x^2 y^{7/3}}{z^3} = \frac{x^2 y^2 (\sqrt[3]{y})}{z^3} \qquad \square$$

Note that $y^{7/3} = y^{(6/3 + 1/3)} = y^{2+1/3} = y^2 y^{1/3} = y^2 (\sqrt[3]{y})$

EXAMPLE 3   **Simplifying expressions with negative exponents**

$$(x^{-1} + y^{-1})^{-1} = \left(\frac{1}{x} + \frac{1}{y}\right)^{-1} = \left(\frac{x+y}{xy}\right)^{-1} = \frac{xy}{x+y} \qquad \square$$

EXAMPLE 4   **Factoring using negative exponents**

$$x^{-\frac{1}{2}} y^{\frac{1}{2}} - x^{\frac{1}{2}} y^{-\frac{1}{2}} = x^{-\frac{1}{2}} y^{-\frac{1}{2}} (y - x) = \frac{y - x}{\sqrt{xy}} \qquad \square$$

EXAMPLE 5   **Expanding an expression using the binomial theorem**

Expand $(2x - 3y)^5$

Solution   In the binomial theorem expansion, replace a by $2x$ and b by $(-3y)$. With $n = 5$, we have

$$(2x - 3y)^5 = (2x)^5 + 5(2x)^4(-3y) + \frac{5 \cdot 4}{2}(2x)^3(-3y)^2$$
$$+ \frac{5 \cdot 4 \cdot 3}{2 \cdot 3}(2x)^2(-3y)^3 + \frac{5 \cdot 4 \cdot 3 \cdot 2}{2 \cdot 3 \cdot 4}(2x)(-3y)^4 + (-3y)^5$$
$$= 32x^5 - 240x^4 y + 720x^3 y^2 - 1080x^2 y^3 + 810xy^4 - 243y^5 \qquad \square$$

2.3

Sequences and Series

Arithmetic sequence   If a_n denotes the nth term, d the common difference, and n the number of terms, then

$$a_n = a_1 + (n - 1)d$$

Arithmetic series   If A_n denotes the nth partial sum of an arithmetic sequence, then

$$A_n = \frac{n}{2}(a_1 + a_n) \text{ or } A_n = \frac{n}{2}[2a_1 + (n - 1)d]$$

Geometric sequence   If g_n denotes the nth term, r the common ratio, and n the number of terms, then

$$g_n = g_1 r^{n-1} \text{ for } n \geq 1$$

Geometric series   If G_n denotes the nth partial sum of a geometric sequence, then

$$G_n = \frac{g_1(1 - r^n)}{1 - r}, \qquad r \neq 1$$

Infinite geometric series   If G is the sum of an infinite geometric series, then

$$G = \frac{g_1}{1 - r}, \quad \text{for } |r| < 1$$

SUMS OF POWERS OF THE FIRST n INTEGERS

$$\sum_{k=1}^{n} 1 = n$$

$$\sum_{k=1}^{n} k = 1 + 2 + 3 + \ldots + n = \frac{n(n+1)}{2}$$

$$\sum_{k=1}^{n} k^2 = 1^2 + 2^2 + 3^2 + \ldots + n^2 = \frac{n(n+1)(2n+1)}{6}$$

$$\sum_{k=1}^{n} k^3 = 1^3 + 2^3 + 3^3 + \ldots + n^3 = \frac{n^2(n+1)^2}{4}$$

$$\sum_{k=1}^{n} k^4 = 1^4 + 2^4 + 3^4 + \ldots + n^4 = \frac{n(n+1)(2n+1)(3n^2+3n-1)}{30}$$

2.4

COMPLETING THE SQUARE

In algebra, the method of completing the square was first introduced as a technique for solving quadratic equations. This method leads to a proof of the quadratic formula. However, when we are working with conic sections, we need to complete the square whenever the conic is not centered at the origin. Consider

$$Ax^2 + Bx + C = 0$$

We wish to put this into the form

$$(x + ?)^2 = \text{some number}$$

To do this, we follow the steps below:

Step 1: Subtract C (the constant term) from both sides.

$$Ax^2 + Bx = -C$$

Step 2: Divide both sides by A ($A \neq 0$; if $A = 0$, the equation would not be quadratic). That is, we want the coefficient of the squared term to be 1.

$$x^2 + \frac{B}{A}x = -\frac{C}{A}$$

Step 3: Add $\left(\frac{B}{2A}\right)^2$ to both sides. That is, take one-half of the coefficient of the first-degree term, square it, and add it to both sides.

$$x^2 + \frac{B}{A}x + \left(\frac{B}{2A}\right)^2 = \frac{B^2}{4A^2} - \frac{C}{A}$$

Step 4: The expression on the left is now a "perfect square" and can be factored.

$$\left(x + \frac{B}{2A}\right)^2 = \frac{B^2}{4A^2} - \frac{C}{A}$$

EXAMPLE 1

Completing the square

Complete the square for $x^2 + 2x - 5 = 0$.

Solution

$$\begin{aligned} x^2 + 2x - 5 &= 0 \\ x^2 + 2x &= 5 \\ x^2 + 2x + 1 &= 5 + 1 \\ (x + 1)^2 &= 6 \end{aligned}$$ □

EXAMPLE 2

Completing the square when the expression involves fractions

Complete the square for $3x^2 + 5x - 4 = 0$.

Solution

$$\begin{aligned} 3x^2 + 5x - 4 &= 0 \\ 3x^2 + 5x &= 4 \\ x^2 + \tfrac{5}{3}x &= \tfrac{4}{3} \\ x^2 + \tfrac{5}{3}x + \left(\tfrac{5}{6}\right)^2 &= \tfrac{4}{3} + \tfrac{25}{36} \\ \left(x + \tfrac{5}{6}\right)^2 &= \tfrac{73}{36} \end{aligned}$$ □

EXAMPLE 3

Completing the square in two variables

Complete the square in both the x and the y terms.

$$x^2 + y^2 + 4x - 9y - 13 = 0$$

Solution

Associate the x and y terms:

$$\begin{aligned} (x^2 + 4x) + (y^2 - 9y) &= 13 \\ (x^2 + 4x + 4) + \left(y^2 - 9y + \tfrac{81}{4}\right) &= 13 + 4 + \tfrac{81}{4} \\ (x + 2)^2 + \left(y - \tfrac{9}{2}\right)^2 &= \tfrac{149}{4} \end{aligned}$$ □

EXAMPLE 4

Completing the square in two variables

Complete the square in both the x and the y terms.

$$2x^2 - 3y^2 - 12x + 6y + 7 = 0$$

Solution   Associate the x and y terms:

$$(2x^2 - 12x) + (-3y^2 + 6y) = -7$$

In this example, we cannot divide by the coefficient of the squared terms, because we need to make *both* the x^2 and the y^2 coefficients equal to one. Instead, we factor

$$2(x^2 - 6x) - 3(y^2 - 2y) = -7$$

Now we complete the square; be sure you add the *same number* to both sides.

$$\underbrace{2(x^2 - 6x + 9)}_{\text{add 18 to both sides}} - \underbrace{3(y^2 - 2y + 1)}_{\text{add } -3 \text{ to both sides}} = -7 + \mathbf{18} - \mathbf{3}$$

$$2(x - 3)^2 - 3(y - 1)^2 = 8 \qquad \qquad \square$$

2.5

SOLVING EQUATIONS

Linear Equations

To solve the first-degree (or linear) equation, isolate the variable on one side. That is, the solution of $ax + b = 0$ is

$$x = -\frac{b}{a}, \quad \text{where} \quad a \neq 0$$

To solve a linear equation, use the following steps.

Step 1: Use the distributive property to clear the equation of parentheses. If it is a rational expression, multiply both sides by the appropriate expression to eliminate the denominator in the problem. Be sure to check the solutions obtained back in the original.

Step 2: Add the same number to both sides of the equality to obtain an equation in which all of the terms involving the variable are on one side and all of the other terms are on the other side.

Step 3: Multiply (or divide) both sides of the equation by the same nonzero number to isolate the variable on one side.

EXAMPLE 1   **Solving a linear equation**

Solve $4(x - 3) + 5x = 5(8 + x)$

Solution

$$\begin{aligned}
4(x - 3) + 5x &= 5(8 + x) \\
4x - 12 + 5x &= 40 + 5x \\
9x - 12 &= 40 + 5x \\
4x - 12 &= 40 \\
4x &= 52 \\
x &= 13
\end{aligned} \qquad \qquad \square$$

EXAMPLE 2   **Solving a rational equation**

Solve $\dfrac{x+1}{x-2} = \dfrac{x+2}{x-2}$

Solution
$$(x+1)(x-2) = (x+2)(x-2)$$
$$x^2 - 2x + x - 2 = x^2 - 4$$
$$-x - 2 = -4$$
$$-x = -2$$
$$x = 2$$

Notice that $x = 2$ causes division by 0, so the solution set is empty. □

EXAMPLE 3   **Solving a literal equation**

Solve $4x + 5xy + 3y^2 = 10$ for x

Solution
$$4x + 5xy = 10 - 3y^2$$
$$(4 + 5y)x = 10 - 3y^2$$
$$x = \frac{10 - 3y^2}{4 + 5y}, \quad y \neq -\frac{4}{5}$$ □

Quadratic Equations

To solve a second-degree (or quadratic) equation, first obtain a zero on one side. Next, try to factor the quadratic. If it is factorable, set each factor equal to zero and solve. If it is not factorable, use the quadratic formula:

THEOREM 2.1   **Quadratic Formula**

The solution of $ax^2 + bx + c = 0$ is
$$x = \frac{-b \pm \sqrt{b^2 - 4ac}}{2a} \quad \text{where} \quad a \neq 0$$

EXAMPLE 4   **Solving a quadratic equation by factoring**

Solve $2x^2 - x - 3 = 0$

Solution   Factoring, $(2x - 3)(x + 1) = 0$. If $2x - 3 = 0$, then $x = \frac{3}{2}$; if $x + 1 = 0$, then $x = -1$. The solution is $x = \frac{3}{2}, -1$.   □

EXAMPLE 5   **Solving a quadratic equation using the quadratic formula**

Solve $5x^2 - 3x - 4 = 0$

Solution $a = 5$, $b = -3$, and $c = -4$; thus (since it does not factor),
$$x = \frac{3 \pm \sqrt{9 - 4(5)(-4)}}{2(5)} = \frac{3 \pm \sqrt{89}}{10}$$ □

EXAMPLE 6	**Solving a literal equation using the quadratic formula**

Solve $x^2 + 2xy + 3y^2 - 4 = 0$ for x

Solution $a = 1$, $b = 2y$, $c = 3y^2 - 4$; thus,

$$x = \frac{-2y \pm \sqrt{4y^2 - 4(1)(3y^2 - 4)}}{2(1)}$$

$$= \frac{-2y \pm \sqrt{16 - 8y^2}}{2}$$

$$= -y \pm \sqrt{4 - 2y^2} \qquad\qquad \square$$

Higher Degree Equations   See Examples 3 and 4 on pp. 33-35.

Trigonometric Equations   See Section 3.7, pp. 46-49.

2.6

SOLVING INEQUALITIES

Linear Inequalities

The first-degree inequality is solved following the steps outlined above for solving first-degree equations. The only difference is that, when we multiply or divide both sides of an inequality by a negative value, the order of the inequality is reversed. That is, if $a < b$, then

$a + c < b + c$	for any number c
$a - c < b - c$	for any number c
$ac < bc$	for any positive number c
$ac > bc$	for any negative number c

A similar result holds if we use $\leq$, $>$, or $\geq$.

Answers to inequality problems are often intervals on a real number line. We use the following notation for intervals:

closed interval (endpoints included): $[a, b]$

open interval (endpoints excluded): (a, b)

(a, ∞)

half-open (or half-closed) interval: $(a, b]$

$[a, b)$

$(-\infty, a]$

To denote an interval that is not connected, we use the notation for the union of two sets:

$$[a, b) \cup (c, d]$$

$$[a, b) \cup (b, c]$$

EXAMPLE 1	**Solving a linear inequality**

Solve $5x + 3 < 3x - 15$

Solution

$$2x + 3 < -15$$
$$2x < -18$$
$$x < -9$$

The solution is $(-\infty, -9)$.   □

EXAMPLE 2	**Solving a linear inequality**

Solve $(2x + 1)(x - 5) \le (2x - 3)(x + 2)$

Solution

$$2x^2 - 9x - 5 \le 2x^2 + x - 6$$

$$-9x - 5 \le x - 6$$

$$-10x \le -1$$

$$x \ge \frac{1}{10}$$

The solution set is $[0.1, \infty)$.   □

The word *between* is often used with a double inequality. We say that **x is between a and b** if $a \le x \le b$ and **x is strictly between a and b** if $a < x < b$.

EXAMPLE 3	**Solving a between relationship**

Solve $-5 \le x + 4 \le 5$

Solution

$$-5 \le x + 4 \le 5$$

$$-5 - 4 \le x + 4 - 4 \le 5 - 4$$

$$-9 \le x \le 1$$

The solution set is $[-9, 1]$.   □

Quadratic Inequalities

The method of solving quadratic inequalities is similar to the method of solving quadratic equations.

Step 1: Obtain a zero on one side of the inequality.

Step 2: Factor, if possible. If the inequality is a rational expression, factor the numerator and the denominator separately.

Step 3: Set each factor equal to zero. These values are not necessarily the solution of the inequality. If the inequality is not factorable, we treat the entire expression as a single factor and solve by the quadratic formula. Values for which the factors are zero are called the *critical values of x.* Plot these values on a number line. The points determine one or more intervals on a number line.

Step 4: Choose some value in each interval. It will make the inequality true or false; we accordingly include or exclude that interval from the solution set.

EXAMPLE 4   **Solving a quadratic inequality by factoring**

Solution   Solve $x^2 < 6 - x$

$$x^2 + x - 6 < 0$$
$$\underbrace{(x - 2)(x + 3)}_{\text{Factors}} < 0$$

Factors < 0 means the product of factors is negative;
factors > 0 means the product of factors is positive.
For this example, we are seeking values of x which make the product of the factors negative.

Signs of factors:

Solve: $x - 2 = 0$
                                 $x = 2$ This is a critical value.
                                           $\downarrow$

$x - 2$: $x < 2 \mid x > 2$
         $x - 2$ is neg $\leftarrow$ $\mid \rightarrow$ **$x - 2$ is pos**
Write: $-\ -\ -\ -\ -\ \mid +\ +\ +\ +\ +$

$x + 3 = 0$
$x = -3$ This is a critical value.
          $\downarrow$

$x + 3$: $x < -3 \mid x > -3$
   $x + 3$ is neg $\leftarrow$ $\mid \rightarrow$ **$x + 3$ is pos**
Write: $-\ -\ -\ \mid +\ +\ +\ +\ +\ +\ +$

We summarize these steps by writing:

$$
\begin{array}{c|c|c}
-\ - & -\ -\ - & +\ +\ +\ +\ + \\
\underbrace{-\ -} & \underbrace{+\ +\ +\ +} & \underbrace{+\ +\ +\ +\ +} \\
\text{neg} \cdot \text{neg} & \text{neg} \cdot \text{pos} & \text{pos} \cdot \text{pos}
\end{array}
$$

$\qquad pos \qquad\qquad neg \qquad\quad pos \qquad \leftarrow product\ of\ factors$

$critical\ values \rightarrow\ -3 \qquad\qquad 2$

We see from the number line that the segment labeled negative is $(-3, 2)$; this is the solution (except for a consideration of the endpoints). In this example, the endpoints are not included, so the solution is an open interval. These steps are summarized as:

$$-\ -\quad -\ +\quad +\ + \qquad \leftarrow signs\ of\ factors$$

$$pos \qquad neg \qquad pos \qquad \leftarrow sign\ of\ product\ (negative\ because\ of\ <\ )$$

Solution: $(-3, 2)$.   □

EXAMPLE 5   **Solving an inequality by examining the factors**

Solve $(2 - x)(x + 3)(x - 1) \geq 0$

Solution   Plotting critical values and checking the signs of the factors, we have

$$+\ -\ -\qquad +\ +\ -\qquad +\ +\ +\qquad -\ +\ +$$

$$pos \qquad neg \qquad pos \qquad neg$$

The solution is $(-\infty, -3] \cup [1, 2]$.   □

EXAMPLE 6   **Solving a rational inequality**

Solve $\dfrac{x + 3}{x - 2} < 0$

Solution   Be careful not to multiply both sides by $(x - 2)$, since you do not know whether $(x - 2)$ is positive or negative. You could consider separate cases, but instead solve it as if it were a quadratic inequality. We set the numerator and denominator each equal to 0 to obtain the critical values $x = -3$, $x = 2$. We then plot these on a number line and check a value in each interval to determine the solution.

$$-\ -\quad +\ -\quad +\ +$$

The solution is $(-3, 2)$.   □

EXAMPLE 7   **Solving a rational inequality**

Solve $\dfrac{x - 3}{x} > 1$

Solution
$$\frac{x - 3}{x} - 1 > 0$$
$$\frac{x - 3 - x}{x} > 0$$
$$\frac{-3}{x} > 0$$

Since $-3 < 0$ for $\dfrac{-3}{x}$ we have:

The solution is $(-\infty, 0)$.   □

EXAMPLE 8

Solve a rational inequality

Solve $\dfrac{x+2}{2x} \geq 5$

Solution

$$\dfrac{x+2}{2x} - 5 \geq 0$$

$$\dfrac{x+2-10x}{2x} \geq 0$$

$$\dfrac{2-9x}{2x} \geq 0$$

The solution is $(0, \frac{2}{9}]$. The endpoints are included when we have intervals with $\geq$ or $\leq$. However, values of the variable that cause division by zero are excluded. □

EXAMPLE 9

Solving a quadratic inequality that does not factor

Solve $x^2 + 2x - 4 < 0$

Solution

The left-hand expression is in simplified form and cannot be factored. Therefore, we proceed by considering $(x^2 + 2x - 4)$ as a single factor. To find the critical values, we find the values for which the factor is 0.

$$x^2 + 2x - 4 = 0$$

$$x = \dfrac{-2 \pm \sqrt{4 - 4(1)(-4)}}{2}$$

$$= -1 \pm \sqrt{5}$$

Plot the critical values, and check the sign of the expression in each of the intervals.

The solution is $(-1 - \sqrt{5}, -1 + \sqrt{5})$. □

2.7

Determinants

The following arrays of numbers are examples of *matrices:*

$$\begin{pmatrix} a_1 & b_1 \\ a_2 & b_2 \end{pmatrix} \qquad \begin{pmatrix} a_1 & b_1 \\ a_2 & b_2 \\ a_3 & b_3 \end{pmatrix} \qquad \begin{pmatrix} a_1 & b_1 & c_1 & d_1 \\ a_2 & b_2 & c_2 & d_2 \end{pmatrix}$$

The first of these matrices has 2 rows and 2 columns, and is called a

"2 by 2 matrix" (written "2×2 matrix"), the second is a 3 × 2 matrix since it has 3 rows and 2 columns, and the last one is a 2 × 4 matrix. The a's, b's, c's, etc. that appear are called the *entries* of the matrix. An $n \times n$ matrix (i.e., one with n rows and n columns, and hence a matrix with the same number of rows and columns) is called a *square* matrix. Associated with each square matrix is a certain number called the *determinant* of the matrix. We shall show how to find the determinant of a 2 × 2 matrix and a 3 × 3 matrix.

DETERMINANT

> If A is the 2 × 2 matrix $\begin{bmatrix} a & b \\ c & d \end{bmatrix}$, then the determinant
>
> of A is defined to be the number $ad - bc$.

Some notations for this determinant are det A, $\begin{vmatrix} a & b \\ c & d \end{vmatrix}$, and $|A|$. We will generally use $|A|$.

EXAMPLE 1

Evaluating determinants

a. $\begin{vmatrix} 4 & -2 \\ -1 & 3 \end{vmatrix} = 4 \cdot 3 - (-2)(-1) = 12 - 2 = 10$

b. $\begin{vmatrix} 2 & 2 \\ 2 & 2 \end{vmatrix} = 2 \cdot 2 - 2 \cdot 2 = 0$

c. $\begin{vmatrix} 1 & 0 \\ 0 & 1 \end{vmatrix} = 1 \cdot 1 - 0 \cdot 0 = 1$

d. $\begin{vmatrix} 1 & 2 \\ 2 & 4 \end{vmatrix} = 1 \cdot 4 - 2 \cdot 2 = 0$ □

We next define the determinant of a 3 × 3 matrix. We need a preliminary observation. If we delete the first row and first column of the matrix

$$A = \begin{bmatrix} a_1 & b_1 & c_1 \\ a_2 & b_2 & c_2 \\ a_3 & b_3 & c_3 \end{bmatrix}$$

We obtain the 2 × 2 matrix $\begin{bmatrix} b_2 & c_2 \\ b_3 & c_3 \end{bmatrix}$. This 2 × 2 matrix is

referred to as the *minor* associated with the entry in the first row and first column. Similarly, the minor associated with b_1 is obtained by deleting the first row and *second* column of A, and hence this minor is

$\begin{bmatrix} a_2 & c_2 \\ a_3 & c_3 \end{bmatrix}$. The minor associated with c_1, is $\begin{bmatrix} a_2 & b_2 \\ a_3 & b_3 \end{bmatrix}$. The

determinant of the 3 × 3 matrix A is obtained using these minors as follows:

$$|A| = a_1 \begin{vmatrix} b_2 & c_2 \\ b_3 & c_3 \end{vmatrix} - b_1 \begin{vmatrix} a_2 & c_2 \\ a_3 & c_3 \end{vmatrix} + c_1 \begin{vmatrix} a_2 & b_2 \\ a_3 & b_3 \end{vmatrix}$$

$$= a_1 \begin{pmatrix} \text{minor associated} \\ \text{with } a_1 \end{pmatrix} - b_1 \begin{pmatrix} \text{minor associated} \\ \text{with } b_1 \end{pmatrix} + c_1 \begin{pmatrix} \text{minor associated} \\ \text{with } c_1 \end{pmatrix}$$

EXAMPLE 2 — Evaluating a determinant

$$\begin{vmatrix} 1 & 4 & -1 \\ -2 & 0 & 2 \\ 3 & 1 & 2 \end{vmatrix} = 1 \begin{vmatrix} 0 & 2 \\ 1 & 2 \end{vmatrix} - 4 \begin{vmatrix} -2 & 2 \\ 3 & 2 \end{vmatrix} + (-1) \begin{vmatrix} -2 & 0 \\ 3 & 1 \end{vmatrix}$$

$$= -2 + 4(10) + 2 = 40$$

It is not difficult to show that $\begin{vmatrix} a & 0 & 0 \\ 0 & b & 0 \\ 0 & 0 & c \end{vmatrix} = abc$ □

We found the determinant for a 3 × 3 matrix by using the entries in the first row and their minors. In fact, it is possible to find the determinant using any row (or column); the formulas are similar and use the minors of the terms in the given row (or column), but adjustments must be made for the signs of the minors. The interested reader can find these results in any standard college algebra book. The following four properties of determinants are worth noting:

THEOREM 2.2 — Properties of determinants

Property 1

If B is the matrix obtained from A by multiplying a row (or column) by a real number k, then $|B| = k|A|$.

Property 2

If B is the matrix obtained from A by adding to a row (or column) of A a multiple of some other row (or column) of A, then $|B| = |A|$.

Property 3

If the matrix B is obtained from A by interchanging two rows (or columns), then $|B| = -|A|$.

Property 4

If two rows (or columns) of a matrix A are equal or proportional, then $|A| = 0$.

EXAMPLE 3 — Properties of determinants illustrated

a. Property 1: $\begin{vmatrix} ka_1 & kb_1 \\ a_2 & b_2 \end{vmatrix} = \begin{vmatrix} a_1 & b_1 \\ ka_2 & kb_2 \end{vmatrix} = k \begin{vmatrix} a_1 & b_1 \\ a_2 & b_2 \end{vmatrix}$

b. Property 2: multiply each entry in the first row of the matrix

$\begin{pmatrix} 1 & 2 \\ 2 & 5 \end{pmatrix} = A$ by -2 and add it to the corresponding entry in the second row, we obtain the matrix $B = \begin{pmatrix} 1 & 2 \\ 0 & 1 \end{pmatrix}$, and property 2 asserts that $|A| = |B|$ (check for yourself to see that this is correct).

c. Property 3 (interchange the first and third rows):

$$\begin{vmatrix} 1 & 3 & -5 \\ 3 & -2 & 1 \\ 7 & 1 & 5 \end{vmatrix} = - \begin{vmatrix} 7 & 1 & 5 \\ 3 & -2 & 1 \\ 1 & 3 & -5 \end{vmatrix}$$

d. $\begin{vmatrix} 1 & 2 \\ 2 & 4 \end{vmatrix} = 0$

Note that 2 times an entry in row 1 is equal to the corresponding entry in row 2.   □

2.8

Functions

Definition: A function f is a rule that assigns to each element x of a set X a unique element y of a set Y. (See Section 1.4 of the text.)

In this book we will simply list and categorize some of the more common types of functions you will encounter in calculus.

ALGEBRAIC FUNCTIONS

Definition: An algebraic function is one that satisfies an equation of the form

$$a_n(x)y^n + a_{n-1}(x)y^{n-1} + \cdots + a_0(x) = 0$$

where the coefficient functions $a_k(x)$ are polynomials in x.

Polynomial Functions

Definition: $f(x) = a_n x^n + a_{n-1}x^{n-1} + a_{n-2}x^{n-2} + \cdots + a_2 x^2 + a_1 x + a_0, \ a_n \neq 0$

Constant function: $f(x) = a$

Linear function: $f(x) = ax + b$
    Standard form: $Ax + By + C = 0$
    Point-slope form: $y - k = m(x - h)$
    Slope-intercept form: $y = mx + b$

$$Two\text{-}point\ form: \qquad y - y_1 = \left(\frac{y_2 - y_1}{x_2 - x_1}\right)(x - x_1) \quad \text{or} \quad \begin{vmatrix} x & y & 1 \\ x_1 & y_1 & 1 \\ x_2 & y_2 & 1 \end{vmatrix} = 0$$

$Intercept\ form: \qquad \frac{x}{a} + \frac{y}{b} = 1$

$Horizontal\ line: \qquad y = k$

$Vertical\ line: \qquad x = h$

Quadratic function: $f(x) = ax^2 + bx + c,\ a \neq 0$

Cubic function: $f(x) = ax^3 + bx^2 + cx + d,\ a \neq 0$

Miscellaneous Algebraic Functions

Absolute value function: $f(x) = |x| = \begin{cases} x & \text{if } x \geq 0 \\ -x & \text{if } x < 0 \end{cases}$

Greatest integer function: $f(x) = [\![x]\!]$ This is the integer $[\![x]\!]$ satisfying $[\![x]\!] \leq x < [\![x]\!] + 1$.

Power function: $f(x) = x^r$ for any real number r.

Rational function: $f(x) = \dfrac{P(x)}{D(x)}$ where $P(x)$ and $D(x)$ are two polynomial functions, $D(x) \neq 0$

TRANSCENDENTAL FUNCTIONS

Definition: Functions that are not algebraic are called transcendental.

Exponential function: $f(x) = b^x$ $(b > 0,\ b \neq 1)$

Logarithmic function: $f(x) = \log_b x$ $(b > 0,\ b \neq 1)$
 If $b = 10$, then f is a common logarithm: $f(x) = \log x$
 If $b = e$, then f is a natural logarithm: $f(x) = \ln x$

Trigonometric functions: Let θ be any angle in standard position and let $P(x,\ y)$ be any point on the terminal side of the angle a distance of r from the origin $(r \neq 0)$. Then $\cos\theta = \frac{x}{r}$, $\sin\theta = \frac{y}{r}$, $\tan\theta = \frac{y}{x}$, $\sec\theta = \frac{r}{x}$, $\csc\theta = \frac{r}{y}$, $\cot\theta = \frac{x}{y}$.

2.9

Polynomials

THEOREMS Let $P(x)$ and $Q(x)$ be polynomial functions.

Zero factor theorem: If $P(x)Q(x) = 0$, then either (or both) $P(x) = 0$ or $Q(x) = 0$.

Remainder theorem: If $P(x)$ is divided by $x - r$ until a constant is obtained, then the remainder is equal to $P(r)$.

Intermediate-value theorem for polynomial function: If $P(x)$ is a polynomial function on $[a, b]$ such that $P(a) \neq P(b)$, then P takes on every value between $P(a)$ and $P(b)$ over the interval $[a, b]$.

Factor theorem: If r is a root of the polynomial equation $P(x) = 0$, then $x - r$ is a factor of $P(x)$. Also, if $x - r$ is a factor of $P(x)$, then r is a root of the polynomial equation $P(x) = 0$.

Root limitation theorem: A polynomial function of degree n has, at most, n distinct roots.

Location theorem: If P is a polynomial function such that $P(a)$ and $P(b)$ are opposite in sign, then there is at least one real root on the interval $[a, b]$.

Rational root theorem: If $P(x)$ has integer coefficients and has a rational root p/q (where p/q is reduced), then p is a factor of the constant term, a_0, and q is a factor of the leading coefficient, a_n.

Upper and lower bound theorem: If $a > 0$ and, in the synthetic division of $P(x)$ by $x - a$, all the numbers in the last row are either positive or negative, then a is an upper bound for the roots of $P(x) = 0$. If $b < 0$ and, in the synthetic division of $P(x)$ by $x - b$, the numbers in the last row alternate in sign, then b is a lower bound for the roots of $P(x) = 0$.

Descartes' rule of signs: Let $P(x)$ be written in descending powers of x.
1. The number of positive real zeros is equal to the number of sign changes or is equal to that number decreased by an even integer.
2. The number of negative real zeros is equal to the number of sign changes in $P(-x)$ or is equal to that number decreased by an even integer.

Fundamental theorem of algebra: If $P(x)$ is a polynomial of degree $n \geq 1$ with complex coefficients, then $P(x) = 0$ has at least one complex root.

Number of roots theorem: If $P(x)$ is a polynomial of degree $n \geq 1$ with complex coefficients, then $P(x) = 0$ has exactly n roots (if roots are counted according to their multiplicity).

EXAMPLE 1	**Synthetic division**

Divide $x^4 + 3x^3 - 12x^2 + 5x - 2$ by $x - 2$

Solution   To do synthetic division, the divisor must be of the form $x - b$.

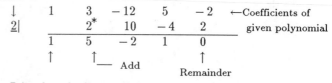

This is the number b.

*The numbers in this row are found as follows:
$$2 \cdot 1 = 2; \ 2 \cdot 5 = 10; \ 2 \cdot (-2) = -4; \ 2 \cdot 1 = 2$$
The degree of the result is one less than the given polynomial and has coefficients given by the last row (1, 5, -2, and 1 for this example; the last entry is the remainder).

$$\frac{x^4 + 3x^3 - 12x^2 + 5x - 2}{x - 2} = x^3 + 5x^2 - 2x + 1 \qquad \square$$

EXAMPLE 2	**Synthetic division with zero coefficients and a remainder**

Divide $x^5 - 3x^2 + 1$ by $x - 2$

Solution

$$
\begin{array}{r|rrrrrr}
  & 1 & 0 & 0 & -3 & 0 & 1 \\
2| &   & 2 & 4 & 8 & 10 & 20 \\
\hline
  & 1 & 2 & 4 & 5 & 10 & 21 \leftarrow (\text{R})
\end{array}
$$

$$\frac{x^5 - 3x^2 + 1}{x - 2} = x^4 + 2x^3 + 4x^2 + 5x + 10 + \frac{21}{x - 2} \qquad \square$$

EXAMPLE 3	**Solving a polynomial equation**

Solve $12x^3 - 6x^2 - 24x + 18 = 0$

Solution

Using the rational root theorem, $p = \pm 1,\ \pm 2,\ \pm 3,\ \pm 6,\ \pm 18$ and $q = 1,\ 2,\ 3,\ 4,\ 6,\ 12$. There are many possible rational roots of the form $\frac{p}{q}$. We try these values using synthetic division until we find a root, which in turn allows us to write a factorization of polynomials of a degree lower than the original polynomial.

$$
\begin{array}{r|rrrr}
  & 12 & -6 & -24 & 18 \\
1 & 12 & 6 & -18 & 0
\end{array}
$$

The root is 1 (zero remainder), so $(x - 1)$ and $(12x^2 + 6x - 18)$ are factors.

$$12x^3 - 6x^2 - 24x + 18 = 0$$
$$(x - 1)(12x^2 + 6x - 18) = 0$$
$$6(x - 1)(2x^2 + x - 3) = 0$$
$$6(x - 1)(2x + 3)(x - 1) = 0$$

The roots are 1, $-\frac{3}{2}$, and 1. $\qquad \square$

EXAMPLE 4	**Solving a polynomial equation with repeated roots**

Solve $r^4 - 5r^3 + 6r^2 + 4r - 8 = 0$

Solution

This equation comes from Example 6, Section 15.2 of the textbook. The possible rational roots are $\pm 1,\ \pm 2,\ \pm 4,$ and ± 8.

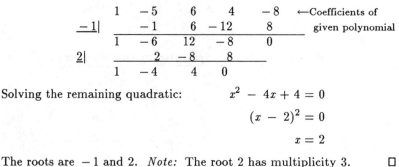

$$
\begin{array}{r}
\;\;\;1 \quad -5 \quad\;\; 6 \quad\;\; 4 \quad -8 \quad \leftarrow\text{Coefficients of} \\
\underline{-1|\quad -1 \quad\;\; 6 \;\; -12 \quad\;\; 8}\quad\; \text{given polynomial} \\
1 \quad -6 \quad 12 \quad -8 \quad\;\; 0 \\
\underline{2|\quad\quad\;\; 2 \quad -8 \quad\;\; 8} \\
1 \quad -4 \quad\;\; 4 \quad\;\; 0
\end{array}
$$

Solving the remaining quadratic: $\qquad x^2 - 4x + 4 = 0$

$$(x - 2)^2 = 0$$

$$x = 2$$

The roots are -1 and 2. *Note:* The root 2 has multiplicity 3. □

| **EXAMPLE 5** | **Solving a polynomial equation which has no rational roots** |

Solve $x^4 - 3x^2 - 6x - 2 = 0$.

Solution

p: ± 1, ± 2 and q: 1, so $\frac{p}{q}$: ± 1, ± 2. We try these values using synthetic division.

$$
\begin{array}{r|rrrrr}
 & 1 & 0 & -3 & -6 & -2 \\
\hline
1 & 1 & 1 & -2 & -8 & -10 \\
2 & 1 & 2 & 1 & -4 & -10 \\
-1 & 1 & -1 & -2 & -4 & 2 \\
-2 & 1 & -2 & 1 & -8 & 14 \leftarrow \text{Lower bound}
\end{array}
$$

There are no rational roots (we have tried all the numbers on our list). Next, verify the type of roots by using Descartes' rule of signs:

$$f(x) = x^4 - 3x^2 - 6x - 2$$

One sign change so there is one positive root.

$$f(-x) = x^4 - 3x^2 + 6x - 2$$

Three sign changes so there are three or one negative roots.

Using synthetic division to find some additional points:

$(0, -2)$ for the y-intercept;

$(1.5, -12.6875)$; $(3, 34)$ is an upper bound.

Using the intermediate value theorem and synthetic division we can find the roots to any desired degree of accuracy. We find

$$
\begin{array}{r|rrrrr}
 & 1 & 0 & -3 & -6 & -2 \\
\hline
-.5 & 1 & -.5 & -2.75 & -4.625 & .3125 \\
-.4 & 1 & -.4 & -2.84 & -4.864 & -.0544 \\
-.41 & 1 & -.41 & -2.8319 & -4.8389 & -.0161 \\
-.42 & 1 & -.42 & -2.8236 & -4.8141 & .0219
\end{array}
$$

Remember that there is root between $-.41$ and $-.42$ since the remainder is positive for $-.42$ and negative for $-.41$. Continue in this fashion to approximate the root to any degree of accuracy desired. This is a good problem for a calculator or a computer. Repeat this procedure for the other roots to find (correct to the nearest tenth): $-.4$, 2.4. The graph can be used to verify that these are the only real roots. $\square$

You may need to consult a precalculus textbook for a review of solving polynomial equations.

2.10 PROBLEM SET 2

A   *Perform the indicated operations in Problems 1-12.*

1. $2^5 \cdot 2^7$ **2.** $5^2 \cdot 5^8$ **3.** $(5^2)^3$

4. $(8^2)^4$ **5.** $\dfrac{3^8}{3^5}$ **6.** $\dfrac{3^5}{3^8}$

7. $2^{-5} \cdot 2^8$ **8.** $\dfrac{2^{-5}}{2^8}$ **9.** $\dfrac{2^{-3}}{2^{-4}}$

10. $8^{2/3} \cdot 4^{1/2}$ **11.** $16^{3/4} \cdot 8^{-1/3}$ **12.** $\dfrac{8^{2/3}}{4^{1/2}}$

Perform the indicated operations (assume a, b, and c are positive real numbers) in Problems 13-22.

13. $(a^2 b^3 c^5) \cdot (a^3 b^5 c^2)$ **14.** $(a^2 b^3 c^5) \cdot (a^{-2} b^{-4} c^{-2})$

15. $(ab^2 c^3)^4$ **16.** $\left(\dfrac{a^2 b^3}{c^2}\right)^5$

17. $\left(\dfrac{a^{-2} b^2}{c^2}\right)^{-1}$ **18.** $\left(\dfrac{a^{-2} b^{-2}}{c^2}\right)^{-1}$

19. $\left(\dfrac{a^{-2} b^{-2}}{c^{-2}}\right)^{-2}$ **20.** $(x^{-2} + y^{-2})$

21. $(x^{-2} + y^{-2})^{-2}$ **22.** $(a^{1/2} b^{2/3} c^{1/5})^{15}$

Factor in Problems 23 - 28.

23. $2a^{1/2} b^{-1/2} + 3a^{-1/2} b^{1/2}$ **24.** $-3a^{-1/2} b + 4a^{1/2}$

25. $a^{1/2} b + a^{-1/2} b^2$ **26.** $a^{1/2} b^{-1/3} + a^{-1/2} b^{1/3}$

27. $(x + 1)^3 (2)(2x + 3)(2) + (2x + 3)^2 (3)(x + 1)^2 (1)$
(From Example 9, Section 3.5 of text.)

28. $4(2x^2 + 1)^3 (4x)(x^2 - 2)^5 + 5(2x^2 + 1)^4 (2x)(x^2 - 2)^4$
(From Problem 23, Section 3.5 of text.)

In Problems 29-54, solve each equation or inequality for x.

29. $3x - 9 \geq 12$ **30.** $-x > -36$ **31.** $3(2 - 4x) \leq 0$

32. $5(3 - x) > 3x - 1$ **33.** $-5 \leq 5x \leq 25$ **34.** $3 \leq -x < 8$

35. $-5 < 3x + 2 \leq 5$ **36.** $-5 \leq 3 - 2x < 18$

37. $x^2 + 5x - 6 = 0$ **38.** $x^2 + 5x + 6 = 0$

39. $3x^2 = 7x$ **40.** $7x^2 = 2$

41. $5x = 3 - 4x^2$ **42.** $4x^2 = 12x - 9$

43. $2x^2 + x - w = 0$ **44.** $2x^2 + wx + 5 = 0$

45. $4x^2 - 4x + (1 - t^2) = 0$ **46.** $y = 2x^2 + x + 6$

47. $4x^2 - (3t + 10)x + (6t + 4) = 0, \; t > 2$

48. $(x - 3)^2 + (y - 2)^2 = 4$

49. $(x + 1)(2x + 5)(7 - 3x) > 0$

50. $(x - 2)(3x + 2)(3 - 2x) < 0$

51. $\dfrac{x(2x - 1)}{5 - x} > 0$ **52.** $\dfrac{x}{(2x + 3)(x - 2)} < 0$

53. $2x^2 + 4x + 5 \geq 0$ **54.** $x^2 - 2x - 6 \leq 0$

In Problems 55 - 66, find the value of the given determinant.

55. $\begin{vmatrix} 3 & 0 & 0 \\ -5 & 2 & 0 \\ \frac{5}{2} & \frac{1}{5} & -1 \end{vmatrix}$ **56.** $\begin{vmatrix} 0 & 1 & 0 \\ -1 & 32 & 1 \\ 2 & 48 & -1 \end{vmatrix}$

57. $\begin{vmatrix} 2 & 0 & 1 \\ 1 & 3 & -2 \\ 2 & -3 & 2 \end{vmatrix}$ **58.** $\begin{vmatrix} 0 & -1 & -1 \\ 1 & 3 & 2 \\ 1 & -4 & 2 \end{vmatrix}$

59. $\begin{vmatrix} -1 & 1 & 2 \\ 0 & 1 & 3 \\ 1 & 0 & -1 \end{vmatrix}$ **60.** $\begin{vmatrix} 2 & -1 & 1 \\ 1 & 0 & 0 \\ 0 & -1 & 2 \end{vmatrix}$

61. $\begin{vmatrix} 1 & -2 & 3 \\ -4 & 7 & -11 \\ 5 & 9 & -1 \end{vmatrix}$ **62.** $\begin{vmatrix} \mathbf{i} & \mathbf{j} & \mathbf{k} \\ 2 & -1 & 3 \\ 0 & 7 & -4 \end{vmatrix}$

(From Example 5   For variables **i**, **j**, and **k**.
Section 10.4 of text.)   (From Example 1,
        Section 10.4 of text.)

63.
$$\begin{vmatrix} 1 & 2 & 3 \\ 0 & 0 & 1 \\ 0 & 1 & 0 \end{vmatrix}$$

64.
$$\begin{vmatrix} 2 & 3 & 4 \\ 2 & 0 & 0 \\ 0 & 1 & 1 \end{vmatrix}$$

65.
$$\begin{vmatrix} 1 & 2 & 3 & 4 \\ 8 & -1 & 5 & 7 \\ 2 & 4 & 6 & 8 \\ -1 & 5 & 3 & 7 \end{vmatrix}$$

66.
$$\begin{vmatrix} 8 & 1 & -7 & 5 \\ -1 & 2 & 2 & 3 \\ -7 & 8 & 15 & 4 \\ -3 & 6 & 6 & 9 \end{vmatrix}$$

In Problems 67-72, expand the given quantity.

67. $(a + b)^4$ **68.** $(a + 2b)^3$ **69.** $(3x + 2y)^3$

70. $(-x + 2y)^5$ **71.** $(\frac{1}{x} + y)^4$ **72.** $(\sqrt{a} + 2b)^5$

Solve the polynomial equations in Problems 73-86.

73. $x^3 - x^2 - 4x + 4 = 0$

74. $2x^3 - x^2 - 18x + 9 = 0$

75. $x^3 - 2x^2 - 9x + 18 = 0$

76. $x^3 + 2x^2 - 5x - 6 = 0$

77. $x^3 + 3x^2 - 4x - 12 = 0$

78. $2x^3 + x^2 - 13x + 6 = 0$

79. $2x^3 - 3x^2 - 32x - 15 = 0$

80. $x^4 - 12x^3 + 54x^2 - 108x + 81 = 0$

81. $x^4 + 3x^3 - 19x^2 - 3x + 18 = 0$

82. $x^4 - 13x^2 + 36 = 0$

83. $x^3 + 15x^2 + 71x + 105 = 0$

84. $x^5 + 6x^4 + x^3 - 48x^2 - 92x - 48 = 0$

85. $x^6 - 3x^4 + 3x^2 - 1 = 0$

86. $x^7 + 3x^6 - 4x^5 - 16x^4 - 13x^3 - 3x^2 = 0$

CHAPTER 3

Review of Trigonometry

3.1

Trigonometric Functions

Let θ be any angle in standard position and let $P(x, y)$ be any point
on the terminal side of the angle a distance of r from the origin
($r \neq 0$). Then

$$\cos \theta = \frac{x}{r} \qquad\qquad \sin \theta = \frac{y}{r} \qquad\qquad \tan \theta = \frac{y}{x}$$

$$\sec \theta = \frac{r}{x} \qquad\qquad \csc \theta = \frac{r}{y} \qquad\qquad \cot \theta = \frac{x}{y}$$

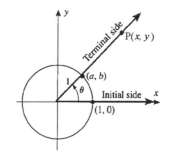

Standard position angle

RADIANS AND DEGREES

$360° = 2\pi$ radians $= 1$ revolution

$$1 \text{ radian} = \left(\frac{180}{\pi}\right)° \approx 57.2957 \cdots \text{ degrees} \qquad 1 \text{ degree} = \left(\frac{\pi}{180}\right) \approx 0.0174532 \cdots \text{ radian}$$

In calculus, it is often necessary to determine when a given function is not defined. For example, in
Problem 17, Problem Set 1.3 in the textbook, for example, we need to know when $f(x) = 3 \tan x - 5$
$\sin x \cos x$ is not defined. From the definition given above, $f(\frac{\pi}{2})$ is not defined since $\tan x$ is not
defined for $x = \frac{\pi}{2}$ (or for any multiple of $\frac{\pi}{2}$).

3.2

Inverse Trigonometric Functions

Inverse Function	Domain	Range
$y = \arccos x$ or $y = \cos^{-1} x$	$-1 \leq x \leq 1$	$0 \leq y \leq \pi$
$y = \arcsin x$ or $y = \sin^{-1} x$	$-1 \leq x \leq 1$	$-\frac{\pi}{2} \leq y \leq \frac{\pi}{2}$
$y = \arctan x$ or $y = \tan^{-1} x$	All reals	$-\frac{\pi}{2} < y < \frac{\pi}{2}$
$y = \text{arccot } x$ or $y = \cot^{-1} x$	All reals	$0 < y < \pi$
$y = \text{arcsec } x$ or $y = \sec^{-1} x$	$|x| \geq 1$	$0 \leq y \leq \pi, y \neq \frac{\pi}{2}$
$y = \text{arccsc } x$ or $y = \csc^{-1} x$	$|x| \geq 1$	$-\frac{\pi}{2} \leq y \leq \frac{\pi}{2}$

The principal values of the inverse trigonometric relations are those values defined by these inverse trigonometric functions. These are the values obtained when using a calculator.

3.3

Evaluating Trigonometric Functions

REDUCTION PRINCIPLE

If t represents any of the six trigonometric functions, then $t(\theta) = \pm t(\theta')$ where the sign depends on the quadrant and θ' is the reference angle of θ.

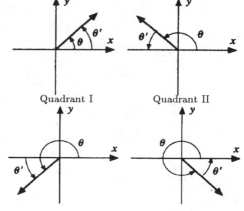

Quadrant I   Quadrant II

Quadrant III   Quadrant IV

All of the trigonometric functions are positive in the first quadrant, sine and cosecant are positive in the second quadrant, tangent and cotangent are positive in the third quadrant, and cosine and secant are positive in the fourth quadrant.

BY CALCULATOR

$\cos \theta$ $\boxed{\theta}$ $\boxed{\cos}$ $\sec \theta$ $\boxed{\theta}$ $\boxed{\sin}$ $\boxed{1/x}$ $\text{Cos}^{-1}\theta$ $\boxed{\theta}$ $\boxed{\text{inv}}$ $\boxed{\cos}$ $\text{Sec}^{-1}\theta$ $\boxed{\theta}$ $\boxed{1/x}$ $\boxed{\text{inv}}$ $\boxed{\cos}$

$\sin \theta$ $\boxed{\theta}$ $\boxed{\sin}$ $\csc \theta$ $\boxed{\theta}$ $\boxed{\sin}$ $\boxed{1/x}$ $\text{Sin}^{-1}\theta$ $\boxed{\theta}$ $\boxed{\text{inv}}$ $\boxed{\sin}$ $\text{Csc}^{-1}\theta$ $\boxed{\theta}$ $\boxed{1/x}$ $\boxed{\text{inv}}$ $\boxed{\sin}$

$\tan \theta$ $\boxed{\theta}$ $\boxed{\tan}$ $\cot \theta$ $\boxed{\theta}$ $\boxed{\tan}$ $\boxed{1/x}$ $\text{Tan}^{-1}\theta$ $\boxed{\theta}$ $\boxed{\text{inv}}$ $\boxed{\tan}$ $\text{Cot}^{-1}\theta$:

if $\theta > 0$; $\boxed{\theta}$ $\boxed{1/x}$ $\boxed{\text{inv}}$ $\boxed{\tan}$

if $\theta < 0$; $\boxed{\theta}$ $\boxed{1/x}$ $\boxed{\text{inv}}$ $\boxed{\tan}$ $\boxed{+}$ $\boxed{\pi}$ $\boxed{=}$

BY TABLE

Angle θ	0	$\frac{\pi}{6}$	$\frac{\pi}{4}$	$\frac{\pi}{3}$	$\frac{\pi}{2}$	π	$\frac{3\pi}{2}$
$\cos\theta$	1	$\frac{\sqrt{3}}{2}$	$\frac{\sqrt{2}}{2}$	$\frac{1}{2}$	0	-1	0
$\sin\theta$	0	$\frac{1}{2}$	$\frac{\sqrt{2}}{2}$	$\frac{\sqrt{3}}{2}$	1	0	-1
$\tan\theta$	0	$\frac{\sqrt{3}}{3}$	1	$\sqrt{3}$	undef.	0	undef.
$\sec\theta$	1	$\frac{2}{\sqrt{3}}$	$\frac{2}{\sqrt{2}}$	2	undef.	-1	undef.
$\csc\theta$	undef.	2	$\frac{2}{\sqrt{2}}$	$\frac{2}{\sqrt{3}}$	1	undef.	-1
$\cot\theta$	undef.	$\sqrt{3}$	1	$\frac{1}{\sqrt{3}}$	0	undef.	0

3.4

Trigonometric Graphs

TRIGONOMETRIC FUNCTIONS

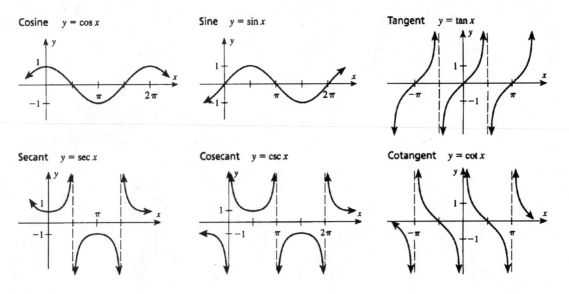

Cosine $y = \cos x$

Sine $y = \sin x$

Tangent $y = \tan x$

Secant $y = \sec x$

Cosecant $y = \csc x$

Cotangent $y = \cot x$

INVERSE TRIGONOMETRIC FUNCTIONS

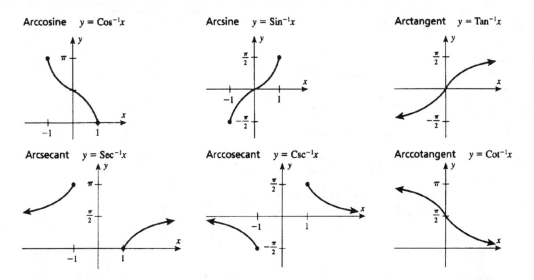

GENERAL TRIGONOMETRIC GRAPHS
To graph a general trigonometric curve, begin by drawing a frame.

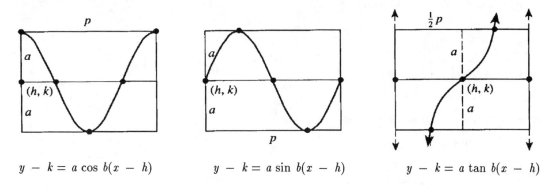

$$y - k = a \cos b(x - h) \qquad y - k = a \sin b(x - h) \qquad y - k = a \tan b(x - h)$$

1.   Algebraically, put the equation in one of the general forms shown above.

2.   Identify (by inspection) the following values: (h, k), a, and b. Then calculate $p = \dfrac{2\pi}{b}$ for the cosine and sine and $p = \dfrac{\pi}{b}$ for the tangent.

3.   Draw a frame:
     a.   Plot (h, k); this is the starting point.
     b.   Draw a; the *height of the frame* is $2|a|$ — up and down a units from (h, k)
     c.   Draw p; the *length of the frame.*

4.   Plot the endpoints, midpoints, and quarter points for each curve as shown in the illustration above. You do not need to know the coordinates of these points. Draw a sketch of the curve using the plotted points.

3.5

Trigonometric Identities

Fundamental identities

Reciprocal identities

1. $\sec \theta = \dfrac{1}{\cos \theta}$ **2.** $\csc \theta = \dfrac{1}{\sin \theta}$ **3.** $\cot \theta = \dfrac{1}{\tan \theta}$

Ratio identities

4. $\tan \theta = \dfrac{\sin \theta}{\cos \theta}$ **5.** $\cot \theta = \dfrac{\cos \theta}{\sin \theta}$

Pythagorean identities

6. $\cos^2 \theta + \sin^2 \theta = 1$ **7.** $1 + \tan^2 \theta = \sec^2 \theta$ **8.** $\cot^2 \theta + 1 = \csc^2 \theta$

Cofunction identities

9. $\cos(\frac{\pi}{2} - \theta) = \sin \theta$ **10.** $\sin(\frac{\pi}{2} - \theta) = \cos \theta$ **11.** $\tan(\frac{\pi}{2} - \theta) = \cot \theta$

Opposite-angle identities

12. $\cos(-\theta) = \cos \theta$ **13.** $\sin(-\theta) = -\sin \theta$ **14.** $\tan(-\theta) = -\tan \theta$

Addition laws

15. $\cos(\alpha + \beta) = \cos \alpha \cos \beta - \sin \alpha \sin \beta$

16. $\cos(\alpha - \beta) = \cos \alpha \cos \beta + \sin \alpha \sin \beta$

17. $\sin(\alpha + \beta) = \sin \alpha \cos \beta + \cos \alpha \sin \beta$

18. $\sin(\alpha - \beta) = \sin \alpha \cos \beta - \cos \alpha \sin \beta$

19. $\tan(\alpha + \beta) = \dfrac{\tan \alpha + \tan \beta}{1 - \tan \alpha \tan \beta}$

20. $\tan(\alpha - \beta) = \dfrac{\tan \alpha - \tan \beta}{1 + \tan \alpha \tan \beta}$

Double-angle identities

21. $\begin{aligned}\cos 2\theta &= \cos^2\theta - \sin^2\theta \\ &= 2\cos^2\theta - 1 \\ &= 1 - 2\sin^2\theta\end{aligned}$ **22.** $\sin 2\theta = 2 \sin \theta \cos \theta$ **23.** $\tan 2\theta = \dfrac{2 \tan \theta}{1 - \tan^2\theta}$

Half-angle identities

24. $\cos \frac{1}{2}\theta = \pm\sqrt{\dfrac{1 + \cos\theta}{2}}$ **25.** $\sin\frac{1}{2}\theta = \pm\sqrt{\dfrac{1 - \cos\theta}{2}}$ **26.** $\tan\frac{1}{2}\theta = \dfrac{1 - \cos\theta}{\sin\theta}$

$$= \dfrac{\sin\theta}{1 + \cos\theta}$$

Product-to-sum identities

27. $2\cos\alpha\cos\beta = \cos(\alpha - \beta) + \cos(\alpha + \beta)$
28. $2\sin\alpha\sin\beta = \cos(\alpha - \beta) - \cos(\alpha + \beta)$
29. $2\sin\alpha\cos\beta = \sin(\alpha + \beta) + \sin(\alpha - \beta)$
30. $2\cos\alpha\sin\beta = \sin(\alpha + \beta) - \sin(\alpha - \beta)$

Sum-to-product identities

31. $\cos\alpha + \cos\beta = 2\cos\left(\dfrac{\alpha + \beta}{2}\right)\cos\left(\dfrac{\alpha - \beta}{2}\right)$

32. $\cos\alpha - \cos\beta = -2\sin\left(\dfrac{\alpha + \beta}{2}\right)\sin\left(\dfrac{\alpha - \beta}{2}\right)$

33. $\sin\alpha + \sin\beta = 2\sin\left(\dfrac{\alpha + \beta}{2}\right)\cos\left(\dfrac{\alpha - \beta}{2}\right)$

34. $\sin\alpha - \sin\beta = 2\sin\left(\dfrac{\alpha - \beta}{2}\right)\cos\left(\dfrac{\alpha + \beta}{2}\right)$

Hyperbolic identities

35. $\operatorname{sech} x = \dfrac{1}{\cosh x}$ **36.** $\operatorname{csch} x = \dfrac{1}{\sinh x}$ **37.** $\coth x = \dfrac{1}{\tanh x}$

38. $\tanh x = \dfrac{\sinh x}{\cosh x}$ **39.** $\coth x = \dfrac{\cosh x}{\sinh x}$

40. $\cosh^2 x - \sinh^2 x = 1$ **41.** $1 - \tanh^2 x = \operatorname{sech}^2 x$ **42.** $\coth^2 x - 1 = \operatorname{csch}^2 x$

43. $\sinh(-x) = -\sinh x$ **44.** $\cosh(-x) = \cosh x$ **45.** $\tanh(-x) = -\tanh x$

46. $\sinh(x \pm y) = \sinh x \cosh y \pm \cosh x \sinh y$

47. $\cosh(x \pm y) = \cosh x \cosh y \pm \sinh x \sinh y$

48. $\tanh(x \pm y) = \dfrac{\tanh x \pm \tanh y}{1 \pm \tanh x \tanh y}$

49. $\cosh 2x = \cosh^2 x + \sinh^2 y$ **50.** $\sinh 2x = 2\sinh x \cosh x$ **51.** $\tanh 2x = \dfrac{2\tanh x}{1 + \tanh^2 x}$

$$= 2\cosh^2 x - 1$$

$$= 1 + 2\sinh^2 x$$

52. $\cosh\frac{1}{2}x = \pm\sqrt{\dfrac{\cosh x + 1}{2}}$ **53.** $\sinh\frac{1}{2}x = \pm\sqrt{\dfrac{\cosh x - 1}{2}}$

54. $\tanh\frac{1}{2}x = \dfrac{\cosh x - 1}{\sinh x} = \dfrac{\sinh x}{\cosh x + 1}$

55. $\sinh^{-1} x = \ln(x + \sqrt{x^2 + 1})$

56. $\operatorname{csch}^{-1} x = \ln\left(\dfrac{1 + \sqrt{1 + x^2}}{x}\right)$ if $x > 0$

57. $\cosh^{-1} x = \ln(x + \sqrt{x^2 - 1}$

58. $\operatorname{sech}^{-1} = \ln\left(\dfrac{1 + \sqrt{1 - x^2}}{\sinh x}\right)$ if $0 < x \leq 1$

3.6

Solving Triangles

The following results hold for any plane triangle ABC with sides a, b, c and angles α, β, γ.

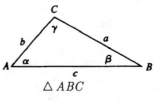

$\triangle\,ABC$

Pythagorean theorem

In a right triangle, $c^2 = a^2 + b^2$

Law of cosines

$a^2 = b^2 + c^2 - 2bc \cos \alpha$ $\qquad$ $b^2 = a^2 + c^2 - 2ac \cos \beta$ $\qquad$ $c^2 = a^2 + b^2 - 2ab \cos \gamma$

$\cos \alpha = \dfrac{b^2 + c^2 - a^2}{2bc}$ $\qquad$ $\cos \beta = \dfrac{a^2 + c^2 - b^2}{2ac}$ $\qquad$ $\cos \gamma = \dfrac{a^2 + b^2 - c^2}{2ab}$

Law of sines

$$\frac{\sin \alpha}{a} = \frac{\sin \beta}{b} = \frac{\sin \gamma}{c}$$

Law of tangents

$$\frac{a + b}{a - b} = \frac{\tan \frac{1}{2}(\alpha + \beta)}{\tan \frac{1}{2}(\alpha - \beta)}$$

with similar relations involving the other sides and angles

AREA FORMULAS

Two sides and an included angle are known:

$$\text{Area} = \tfrac{1}{2}ab \sin \gamma = \tfrac{1}{2}ac \sin \beta = \tfrac{1}{2}bc \sin \alpha$$

Two angles and an included side are known:

$$\text{Area} = \frac{a^2 \sin \beta \sin \gamma}{2 \sin \alpha} = \frac{b^2 \sin \alpha \sin \gamma}{2 \sin \beta} = \frac{c^2 \sin \alpha \sin \beta}{2 \sin \gamma}$$

Three sides are known:

$$\text{Area} = \sqrt{s(s - a)(s - b)(s - c)} \text{ where } s = \tfrac{1}{2}(a + b + c)$$

Area of a sector

$$\text{Area} = \tfrac{1}{2}\theta r^2$$

PROCEDURE FOR SOLVING TRIANGLES

Given	Conditions on the given information	Method of Solution
1. AAA		No solution
2. SSS	**More than one side given:** **a.** The sum of the lengths of the two smaller sides is less than or equal to the length of the larger side. **b.** The sum of the lengths of the two smaller sides is greater than the length of the larger side.	**Law of Cosines** No solution Law of Cosines
3. SAS	**a.** The given angle is greater than or equal to 180°. **b.** The given angle is less than 180°.	No solution Law of Cosines
3. SSA	**a.** The given angle is greater than or equal to 180°. **b.** There are no solutions, one solution, or two solutions as determined by the quadratic formula.	No solution Law of Cosines
4. ASA or AAS	**More than one angle given:** **a.** The sum of the given angles is greater than or equal to 180°. **b.** The sum of the given angles is less than 180°.	**Law of Sines** No solution **Law of Sines**

POLAR-FORM POINTS

To plot a point (r, θ) in polar form, rotate the polar axis through an angle θ, as shown in Figure 3.1. You may find it helpful to think of the polar axis as a pencil — the tip points in the positive direction, and the eraser in the negative direction. After the polar axis is rotated, then plot the distance r on the polar axis (pencil). Note the positive r-values are in the direction of the tip and negative r-values are in the direction of the eraser.

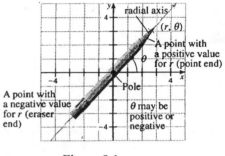

Ordered pairs in polar form are not associated in a one-to-one fashion with points in the plane. Any given point in the plane has infinitely many ordered pairs of polar coordinates associated with that point. If the point (r, θ)

Figure 3.1

other than the pole is given in polar coordinates, then $(-r, \theta + \pi)$ also represents that same point. In addition, there are infinitely many others, all of which have the same first component as one of these representations and have second components that are multiples of 2π added to these angles. The forms (r, θ) and $(-r, \theta + \pi)$ are the **primary representations of the points** if the angles θ and $\theta + \pi$ are between 0 and 2π.

3.7
Trigonometric Equations

PROCEDURE FOR
SOLVING
TRIGONOMETRIC
EQUATIONS

1. Solve for a single trigonometric function. You may use trigonometric identities, factoring, or the quadratic formula.
2. Solve for the angle. You will use the definition of the inverse trigonometric functions for this step.
3. Solve for the unknown.

PRINCIPAL VALUES

EXAMPLE 1

Linear trigonometric equation

Solve the equations for the principle value of θ. Compare the solutions of these similar problems.

a. $5 \sin(\theta + 1) = \frac{1}{2}$ **b.** $\sin 5(\theta + 1) = \frac{1}{2}$

Solution   **a.** $5 \sin(\theta + 1) = \frac{1}{2}$

$$\sin(\theta + 1) = \frac{1}{10} \qquad \textit{Solve for the function}$$

$$\theta + 1 = \sin^{-1} 0.1 \qquad \textit{Solve for the angle}$$

$$\theta = -1 + \sin^{-1} 0.1 \qquad \textit{Solve for the unknown}$$

$$\approx -0.8998326 \qquad \textit{We use radians (real numbers) unless otherwise specified}$$

b. $\sin 5(\theta + 1) = \frac{1}{2}$ *It is solved for the function*

$$5(\theta + 1) = \sin^{-1}\left(\tfrac{1}{2}\right) \qquad \textit{Solve for the angle}$$

$$5\theta = -5 + \frac{\pi}{6} \qquad \textit{Solve for the unknown}$$

$$= -1 + \frac{\pi}{30}$$

$$\approx -0.8952802$$

Notice that these two answers are *not* the same.   □

_____

EXAMPLE 2   **Quadratic trigonometric equation by factoring**
_____

Find the principal values of θ for $15 \cos^2\theta - 2 \cos\theta - 8 = 0$.

Solution $15 \cos^2\theta - 2 \cos\theta - 8 = 0$

$$(3 \cos\theta + 2)(5 \cos\theta - 4) = 0 \qquad \textit{Factor, if possible}$$

This is solved by setting each factor equal to zero (factor theorem):

$3 \cos\theta + 2 = 0$	$5 \cos\theta - 4 = 0$
$\cos\theta = -\frac{2}{3}$	$\cos\theta = \frac{4}{5}$
$\theta = \cos^{-1}\left(-\frac{2}{3}\right)$	$\theta = \cos^{-1}\left(\frac{4}{5}\right)$
≈ 2.300524	≈ 0.6435011 □

_____

EXAMPLE 3   **Quadratic trigonometric equation by quadratic formula**
_____

Find the principal values of θ for $\tan^2\theta - 5 \tan\theta - 4 = 0$.

Solution $\tan^2\theta - 5 \tan\theta - 4 = 0$ *Factor, if possible*

$$\tan\theta = \frac{5 \pm \sqrt{25 - 4(1)(-4)}}{2}$$

Quadratic formula if it cannot be factored

$$= \frac{5 \pm \sqrt{41}}{2}$$

$$\theta = \tan^{-1}\left(\frac{5 \pm \sqrt{41}}{2}\right)$$

$$\approx \tan^{-1}(5.7015621),\ \tan^{-1}(-0.7015621)$$

$$\approx 1.3971718,\ -0.6117736 \qquad \square$$

GENERAL SOLUTION

The procedure for finding the general solution of a trigonometric equation is to first find the reference angle. It can be found by using the table of exact values or a calculator. Then find two values less than one revolution using the reference angle:

For $y = \arccos x$ or $y = \operatorname{arcsec} x$;
  if y is positive, then the inverse function is in Quadrants I and IV
  if y is negative, then the inverse function is in Quadrants II and III

Sine and cosecant positive	**A**ll positive
Tangent and cotangent positive	**C**osine and secant positive

Figure 3.2 Signs of the trigonometric functions

For $y = \arcsin x$ or $y = \operatorname{arccsc} x$;
  if y is positive, then the inverse function is in Quadrants I and II
  if y is negative, then the inverse function is in Quadrants III and IV

For $y = \arctan x$ or $y = \operatorname{arccot} x$;
  if y is positive, then the inverse function is in Quadrants I and III
  if y is negative, then the inverse function is in Quadrants II and IV

The above information is summarized in Figure 3.2.

After you have found the two values less than one revolution, the entire solution is found by using the period of the function:

For cosine, sine, secant and cosecant: add multiples of 2π
For tangent and cotangent: add multiples of π

EXAMPLE 4

General linear trigonometric equation

a. Solve $\cos\theta = \frac{1}{2}$ **b.** Solve $\cos\theta = -\frac{1}{2}$

Solution

a. Reference angle $\theta' = \operatorname{Arccos}\left|\frac{1}{2}\right| = \frac{\pi}{3}$; Angles less than one revolution with a reference angle of $\frac{\pi}{3}$ are $\frac{\pi}{3}$ (Quadrant I) and $\frac{5\pi}{3}$ (Quadrant IV). The solution is infinite, so to find all solution add multiples of 2π to these values: $\theta = \frac{\pi}{3} + 2k\pi$, $\frac{5\pi}{3} + 2k\pi$ for any integral value of k.

b. $\theta' = \operatorname{Arccos}\left|-\frac{1}{2}\right| = \frac{\pi}{3}$. Reference angle is in Quadrants II and III: $\theta = \frac{2\pi}{3} + 2k\pi, \frac{4\pi}{3} + 2k\pi$. $\qquad \square$

EXAMPLE 5

Approximate values of a linear trigonometric function

Solve $\sin x = \frac{2}{\pi}$ for $0 \le x < 2\pi$

Solution

$x = \operatorname{Arcsin}\left(\frac{2}{\pi}\right) \approx 0.6366197$. The solution is in Quadrants I and II ($\frac{2}{\pi}$ is positive): $x \approx 0.6901071, 2.4514856$. $\qquad \square$

| **EXAMPLE 6** | **Trigonometric equation using an identity** |

Solve $-\frac{1}{2}\cos 2t = -2\sin t$

Solution

$$-\frac{1}{2}\cos 2t = -2\sin t$$

$$\cos 2t = 4\sin t$$

$$1 - 2\sin^2 t = 4\sin t$$

$$2\sin^2 t + 4\sin t - 1 = 0$$

$$\sin t = \frac{-4 \pm \sqrt{4^2 - 4(2)(-1)}}{2(2)}$$

$$= \frac{-4 \pm 2\sqrt{6}}{4}$$

$$= \frac{-2 \pm \sqrt{6}}{2}$$

$$\approx 0.2247448714, \ \underset{\uparrow}{-2.224744871}$$

$$\text{Reject } (\sin t \geq -1)$$

$$t \approx \text{Arcsin}(0.227448714)$$

$$\approx 0.2266812, \ 2.914911452 \qquad \square$$

3.8 PROBLEM SET 3

A *Find the radian measure of each angle in Problems 1-4.*

1. $\theta = 6°$ **2.** $\theta = 270°$ **3.** $\theta = 100°$ **4.** $\theta = 2\pi°$

Find the degree measure of each angle in Problems 5-8.

5. $\theta = \frac{2\pi}{3}$ **6.** $\theta = \frac{n\pi}{4}$ **7.** $\theta = 4.2$ **8.** $\theta = 120$

9. Evaluate: **a.** $\cos \pi$ **b.** $\cos \frac{3\pi}{2}$

  c. $\cos 2\pi$ **d.** $\sin \pi$ **e.** $\sin \frac{3\pi}{2}$

10. Evaluate: **a.** $\sin \frac{3\pi}{4}$ **b.** $\cos \frac{3\pi}{4}$

  c. $\sin \frac{5\pi}{4}$ **d.** $\cos \frac{5\pi}{4}$ **e.** $\sin 2\pi$

11. Find $\sin^2\left(\frac{7\pi x}{8}\right) + \cos^2\left(\frac{7\pi x}{8}\right)$

Graph the functions in Problems 16-21.

12. $f(x) = \sin\left(x - \frac{\pi}{3}\right)$ **13.** $f(x) = 3\sin x$

14. $y - 2 = \sin\left(x - \frac{\pi}{2}\right)$ **15.** $y - 1 = 2\cos\left(x - \frac{\pi}{4}\right)$

16. $f(x) = 2\cos(3x + 2\pi) - 2$ **17.** $f(x) = \tan\left(x + \frac{\pi}{4}\right) - 2$

18. $y - 3 = \sin(\pi - x)$ **19.** $y + 1 = -\tan\left(x - \frac{\pi}{6}\right)$

20. $f(x) = 2\sec(x + 1)$ **21.** $f(x) = 2\cot\left(x - \frac{\pi}{6}\right) - 1$

Use the addition formula in Problems 22-25 to find the exact value.

22. $\cos(75°) = \cos(45° + 30°)$ **23.** $\sin 15°$

24. $\sin(105°) = \sin(60° + 45°)$ **25.** $\sin 195°$

26. Show that $\sin 2x = 2 \sin x \cos x$

27. Show that for any integer n, $\sin(x + n\pi) = (-1)^n \sin x$

28. Show that for any integer n, $\cos(x + n\pi) = (-1)^n \cos x$

29. Show that $2 \sin x \cos y = \sin(x + y) + \sin(x - y)$

30. Show that $2 \cos x \sin y = \sin(x + y) - \sin(x - y)$

31. Use the result of Problem 29 to show that
$$\sin A + \sin B = 2 \sin\left(\frac{A + B}{2}\right)\cos\left(\frac{A - B}{2}\right)$$

32. If $\sin^{-1}\frac{1}{5}$ is the angle whose sine is $\frac{1}{5}$ and $\cos^{-1}\frac{1}{5}$ is the angle whose cosine is $\frac{1}{5}$, find the exact value of
$$\cos(\sin^{-1}\tfrac{1}{5} + 2\cos^{-1}\tfrac{1}{5})$$
(From Problem Set 1.5, Problem 29 of text.)

33. If $\sin^{-1}\frac{1}{5}$ is the angle whose sine is $\frac{1}{5}$ and $\cos^{-1}\frac{1}{4}$ is the angle whose cosine is $\frac{1}{4}$, find the exact value of
$$\sin(\sin^{-1}\tfrac{1}{5} + \cos^{-1}\tfrac{1}{4})$$
(From Problem Set 1.5, Problem 30 of text.)

Solve the equations in Problems 34-52 for $0 \leq x < 2\pi$.   Give answers in Problems 42-52 correct to two decimal places.

34. $\sin x = .5$ **35.** $\sin x = -.5$

36. $(\sin x)(\cos x) = 0$ **37.** $(\sin x)(\tan x) = 0$

38. $(2 \cos x + \sqrt{2})(2 \cos x - 1) = 0$

39. $(3 \tan x + \sqrt{3})(3 \tan x - \sqrt{3}) = 0$

40. $\tan^2 x = \tan x$ **41.** $\tan^2 x = \sqrt{3} \tan x$

42. $\cos^2 x - 1 - \cos x = 0$ **43.** $\sin^2 x - \sin x - 2 = 0$

44. $\tan^2 x - 3 \tan x + 1 = 0$ **45.** $\csc^2 x - \csc x - 1 = 0$

46. $\cos 3x + 2 \sin 2x \cos 3x = 0$ **47.** $\sin 2x + 2 \cos x \sin 2x = 0$

48. $\cos(3x - 1) = \frac{1}{2}$ **49.** $\tan(2x + 1) = \sqrt{3}$

50. $\sin 2x + 1 = \sqrt{3}$

51. $\sin^2 3x + \sin 3x + 1 = 1 - \sin^2 3x$

52. $\sin^2 3x + \sin 3x = \cos^2 3x - 1$

CHAPTER 4

4.1

Conic Sections: Parabolas

Consider the general second-degree equation

$$Ax^2 + Bxy + Cy^2 + Dx + Ey + F = 0$$

for any constants A, B, C, D, E, and F. If $A = B = C = 0$, the equation is not quadratic but linear (first degree); but if at least one of A, B, or C is not zero, then the equation is quadratic. Historically, second-degree equations in two variables were first considered in a geometric context and were called **conic sections** because the curves they represent can be described as the intersections of a double-napped right circular cone and a plane.

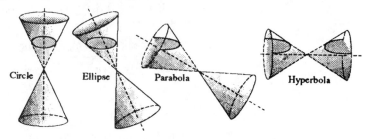

Circle   Ellipse   Parabola   Hyperbola

Figure 4.1 Conic sections

STANDARD-POSITION PARABOLAS

We shall use the following definition for a parabola.

PARABOLA

> A **parabola** is the set of all points in the plane that are equidistant from a fixed point (called the **focus**) and a fixed line (called the **directrix**).

The line through the focus perpendicular to the directrix is called the **principal axis** of the parabola, and the point where the axis intersects the parabola is called the **vertex**. The line segment that passes through the focus perpendicular to the axis and with endpoints on the parabola is called the **focal chord**. This terminology is shown in Figure 4.2.

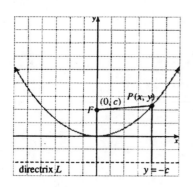

Figure 4.3 Graph of the parabola $x^2 = 4cy$

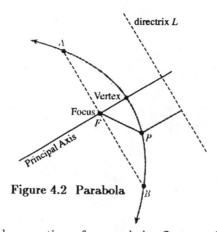

Figure 4.2 Parabola

To obtain the equation of a parabola, first consider a special case: a parabola with focus $F(0, c)$ and directrix $y = -c$, where c is any positive number. This parabola must have its vertex at the origin (the vertex is halfway between the focus and the directrix) and must open upward, as shown in Figure 4.3.

Let (x, y) be any point on the parabola. Then, from the definition of a parabola,

DISTANCE FROM (x, y) to $(0, c)$ = DISTANCE FROM (x, y) TO DIRECTRIX

$$\sqrt{(x - 0)^2 + (y - c)^2} = y + c$$
$$x^2 + y^2 - 2cy + c^2 = y^2 + 2cy + c^2 \qquad \textit{Square both sides.}$$
$$x^2 = 4cy$$

This is the equation of the parabola with vertex $(0, 0)$ and directrix $y = -c$.

You can repeat this argument (see Problems 54-56) for parabolas that have their vertex at the origin and open downward, to the left, and to the right to obtain the results summarized next.

STANDARD-FORM
EQUATIONS FOR
PARABOLAS

Parabola	Focus	Directrix	Vertex
Upward: $x^2 = 4cy$	$(0, c)$	$y = -c$	$(0, 0)$
Downward: $x^2 = -4cy$	$(0, -c)$	$y = c$	$(0, 0)$
Right: $y^2 = 4cx$	$(c, 0)$	$x = -c$	$(0, 0)$
Left: $y^2 = -4cx$	$(-c, 0)$	$x = c$	$(0, 0)$

The length of the *focal chord* is the coefficient $4c$.

To graph a parabola, find and plot the vertex, determine c (usually by inspection), and count out c units from the vertex *in the appropriate direction* as determined by the form of the equation. Finally, it is shown in the problem set that the length of the focal chord is $4c$, and we use this number to determine the width of the parabola, as shown in the following example.

| EXAMPLE 1 | **Graphing a standard-form parabola** |

Graph $2y^2 - 5x = 0$.

Solution First, algebraically change the equation so that it is in standard form by solving for the second-degree term:
$$y^2 = \tfrac{5}{2}x$$
The vertex is $(0, 0)$ and
$$4c = \tfrac{5}{2} \qquad \text{so} \qquad c = \tfrac{5}{8}$$
Thus, the parabola opens to the right, the focus is $(\tfrac{5}{8}, 0)$, and the length of the focal chord is $4c = \tfrac{5}{2}$, as shown in Figure 4.4.

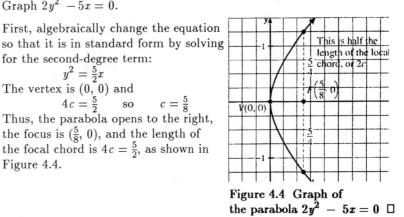

Figure 4.4 Graph of the parabola $2y^2 - 5x = 0$ □

There are two basic types of problems in analytic geometry:

1. Given the equation, draw the graph; this is what we did in Example 1.
2. Given the graph (or information about the graph), write the equation. The next example is of this type.

| EXAMPLE 2 | **Writing the equation of a parabola** |

Find an equation of a parabola with focus $F(0, -2)$ and directrix $y = 2$.

Solution This is the curve drawn in Figure 4.5. We see this is a parabola that opens downward with vertex at the origin. By inspection, the value of c is $c = 2$. The form of the equation is $x^2 = -4cy$, so the desired equation is

$$x^2 = -8y$$

Figure 4.5 Graph of the parabola with focus $(0, -2)$ and directrix $y = 2$

4.2

TRANSLATION OF PARABOLAS

If a parabola is not in standard position but its axis is parallel to one of the coordinate axes, it can be put into standard form by a change of variable of the form $X = x - h$, $Y = y - k$. Such a change in variable is called a **translation**, and it has the general effect shown in Figure 4.6.

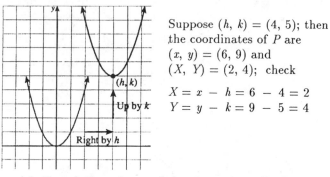

Suppose $(h, k) = (4, 5)$; then the coordinates of P are $(x, y) = (6, 9)$ and $(X, Y) = (2, 4)$; check

$X = x - h = 6 - 4 = 2$
$Y = y - k = 9 - 5 = 4$

Figure 4.6 Translation of a parabola; vertex is at (h, k).

EFFECT OF A
TRANSLATION

Replacing $x - h$ by X and $y - k$ by Y in an equation has the effect of translating the graph of an equation

h units horizontally (**right** if $h > 0$ and **left** if $h < 0$)

k units vertically (**up** if $k > 0$ and **down** if $k < 0$)

This means the equations for parabolas with **vertex** (h, k) are:

$$(x - h)^2 = 4c(y - k), \qquad X^2 = 4cY$$
$$(x - h)^2 = -4c(y - k), \qquad X^2 = -4cY$$
$$(y - k)^2 = 4c(x - h), \qquad Y^2 = 4cX$$
$$(y - k)^2 = -4c(x - h), \qquad Y^2 = -4cX$$

EXAMPLE 1

Graphing a parabola using a translation

Sketch the parabola $y = x^2 + 2x + 3$. Find the vertex, c, and the length of the focal chord. Also, find the focus and the equation of the directrix.

Solution   First, complete the square:

$$y = x^2 + 2x + 3$$

$$y - 3 \boxed{+1} = x^2 + 2x \boxed{+1}$$

$$y - 2 = (x + 1)^2$$

Next, plot the vertex $(-1, 2)$. If we replace $y - 2$ by Y and $x + 1$ by X, we have $Y = X^2$, which tells us the parabola opens upward, $4c = 1$, and $c = \frac{1}{4}$. Thus, plot the focus by counting up $\frac{1}{4}$ unit, and then draw the focal chord with length 1. Because these points are fairly close on the scale we have chosen, we plot an additional point, say the y-intercept: if $x = 0$, then $y = 0^2 + 2(0) + 3 = 3$. See Figure 4.7.

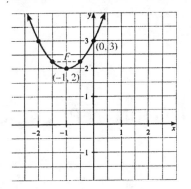

Figure 4.7 Graph of $y = x^2 + 2x + 3$

Notice that we do not need to know the coordinates of the focus or the directrix to draw the graph. We needed to know only the vertex, the distance c, and the length of the focal chord $4c$. If c is small relative to the vertex, it may be necessary to plot an additional point.

However, we may need to know the coordinates of the focus and directrix for further analysis. To find these numbers, we can use the reverse translation $x = X + h$, $y = Y + k$.

	XY-coordinates	xy-coordinates
Vertex	$(0, 0)$	$(-1, 2)$
focus	$(0, \frac{1}{4})$	$(-1, \frac{9}{4})$
directrix	$Y = -\frac{1}{4}$	$y = \frac{7}{4}$

□

EXAMPLE 2   **Finding the equation of a translated parabola**

Find an equation for the parabola with focus $(4, -3)$ and directrix the line $x + 2 = 0$.

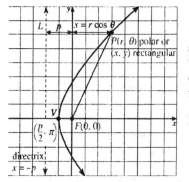

Solution

Figure 4.8 Graph of a parabola with focus $(4, -3)$ **and directrix** $x + 2 = 0$.

Sketch the information as shown in Figure 4.8. The vertex is $(1, -3)$, because it must be equidistant from F and the directrix. Note that $c = 3$. Thus, substitute into the equation of a parabola that opens to the right namely $y^2 = 4cx$, and then translate to the point (h, k) to obtain the equation

$$(y - k)^2 = 4c(x - h)$$

The desired equation, because $(h, k) = (1, -3)$, is

$$(y + 3)^2 = 12(x - 1)$$

□

4.3

REPRESENTATION IN POLAR COORDINATES

Figure 4.9 A standard

polar position for a parabola

Next, we shall see how a parabola can be represented in polar coordinates. Let a parabola be given in the plane. Place the x-axis along the principal axis of the parabola, and place the pole at the focus, as shown in Figure 4.9. We refer to this as a **standard polar position** for the parabola. Assume the parabola opens to the right and that the directrix L is the vertical line $x = -p$, where $p > 0$ is the distance from the focus to the directrix.

If P is a point on the parabola with rectangular coordinates (x, y) and polar coordinates (r, θ), we must have

DISTANCE P TO F = DISTANCE P TO L

$$|r| = |p + r \cos \theta|$$

$$r = \pm(p + r \cos \theta)$$

$$r = \frac{p}{1 - \cos \theta} \qquad -r = \frac{p}{1 + \cos \theta}$$

It can be shown that these two equations represent the same graph, so we shall use the one on the left to represent the given parabola. We can similarly derive equations for parabolas that open downward, left, or to the right. These graphs are shown in Figure 4.10.

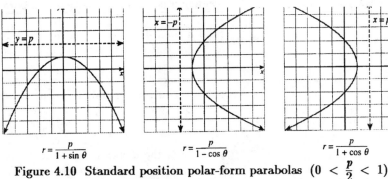

$$r = \frac{p}{1 - \sin \theta} \qquad r = \frac{p}{1 + \sin \theta} \qquad r = \frac{p}{1 - \cos \theta} \qquad r = \frac{p}{1 + \cos \theta}$$

Figure 4.10 Standard position polar-form parabolas $\left(0 < \dfrac{p}{2} < 1\right)$

STANDARD POLAR
EQUATIONS FOR
PARABOLAS

	Parabola	Focus	Directrix Rect-form*	Polar-form Vertex
Upward:	$r = \dfrac{p}{1 - \sin \theta}$	$(0, 0)$	$y = -p$	$\left(\dfrac{p}{2}, \dfrac{3\pi}{2}\right)$
Downward:	$r = \dfrac{p}{1 + \sin \theta}$	$(0, 0)$	$y = p$	$\left(\dfrac{p}{2}, \dfrac{\pi}{2}\right)$
Right:	$r = \dfrac{p}{1 - \cos \theta}$	$(0, 0)$	$x = -p$	$\left(\dfrac{p}{2}, \pi\right)$
Left:	$r = \dfrac{p}{1 + \cos \theta}$	$(0, 0)$	$x = p$	$\left(\dfrac{p}{2}, 0\right)$

EXAMPLE 1

Graphing a polar-form parabola

Describe and sketch the graph of the equation

$$r = \frac{4}{3 - 3 \cos \theta}$$

Solution

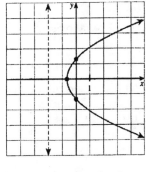

$$r = \frac{4}{3 - 3 \cos \theta} \cdot \frac{\frac{1}{3}}{\frac{1}{3}} = \frac{\frac{4}{3}}{1 - \cos \theta}$$

By inspection, you can now see (Figure 4.11) that the parabola opens to the right and that $p = \frac{4}{3}$. Thus the vertex is

$$\left(\frac{p}{2}, \pi\right) = \left(\frac{2}{3}, \pi\right)$$

Plot the vertex and the line $x = -\frac{4}{3}$, as shown in the margin. You can plot other points that are easy to calculate, such as the points where $\theta = \frac{\pi}{2}$ and $\theta = \pi$ or $\theta = \frac{3\pi}{2}$.   □

Figure 4.11   Graph of
$$r = \frac{4}{3 - 3 \cos \theta}$$

---

*We could, of course, state these equations of lines in polar form. For example, $y = -p$ is $r \sin \theta = -p$, but we prefer writing the equation of the directrix in rectangular form.

EXAMPLE 2   **Finding a polar-form equation of a parabola**

Find a polar-form equation for the parabola with focus at the origin and vertex at the polar-form point $(3, \pi)$.

Solution   The vertex $(3, \pi)$ is on the x-axis and to the left of the focus (the pole). Thus, the parabola opens to the right and has a polar-form equation

$$r = \frac{p}{1 - \cos\theta}$$

where p is the distance from the focus to the directrix. Because the vertex $(3, \pi)$ is halfway between the focus and the directrix, we must have $p = 6$, so that the required equation is '

$$r = \frac{6}{1 - \cos\theta} \qquad \Box$$

4.4

PARABOLIC REFLECTORS*

Parabolic curves are used in the design of lighting systems, telescopes, and radar antennas, mainly because of the property illustrated in Figure 4.12 and described more formally in Theorem 4.3.

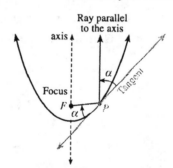

Figure 4.12 The reflection property of parabolas

THEOREM 4.3   **Reflection property of parabolas**

Let P be a point on a parabola in the plane, and let T be the tangent line to the parabola at P. Then the angle between T and the line through P parallel to the principal axis of the parabola equals the angle between T and the line connecting P to the focus.

---

*Since these reflectors are three dimensional, the precise word is *paraboloidal*, but the most common usage refers to the cross-sectional shape which is parabolic.

As an illustration of how this property is used, let us examine its application to reflecting telescopes. The eyepiece of such a telescope is placed at the focus of a parabolic mirror. Light enters the telescope in rays that are parallel to the axis of the parabola. It is a principle of physics that when light is reflected, the angle of incidence equals the angle of reflection. Hence, the parallel rays of light strike the parabolic mirror so that they all reflect through the focus, which means that the parallel rays are concentrated at the eyepiece located at the focus.

Flashlights and automobile headlights simply reverse the process: A light source is placed at the focus of a parabolic mirror, the light rays strike the mirror with angle of incidence equal to the angle of reflection, and each ray is reflected along a path parallel to the axis, thus emitting a light beam of parallel rays.

Radar utilizes both of these properties. First, a pulse is transmitted from the focus to a parabolic surface. As with a reflecting telescope, parallel pulses are transmitted in this way. The reflected pulses then strike the parabolic surface and are sent back to be received at the focus.

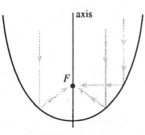

A reflecting telescope: Light rays parallel to the axis are concentrated at the focus.

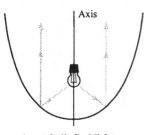

A parabolic flashlight: A light source at the focus sends out beams of light parallel to the axis.

4.5 PROBLEM SET 4

A   *Sketch the curves in Problems 1-22. Find the vertex, V, and c.*

1. $y^2 = 8x$

2. $y^2 = -12x$

3. $y^2 = -20x$

4. $4x^2 = 10y$

5. $3x^2 = -12y$

6. $2x^2 = -4y$

7. $2x^2 + 5y = 0$

8. $5y^2 + 15x = 0$

9. $3y^2 - 15x = 0$

10. $4y^2 + 3x = 12$

11. $5x^2 + 4y = 20$

12. $4x^2 + 3y = 12$

13. $(y - 1)^2 = 2(x + 2)$

14. $(y + 3)^2 = 3(x - 1)$

15. $(x + 2)^2 = 2(y - 1)$

16. $(x - 1)^2 = 3(y + 3)$

17. $y^2 + 4x - 3y + 1 = 0$

18. $y^2 - 4x + 10y + 13 = 0$

19. $y^2 + 4y - 10x + 74 = 0$

20. $x^2 + 9y - 6x + 18 = 0$

21. $9x^2 + 6x + 18y - 23 = 0$

22. $9x^2 + 6y + 18x - 23 = 0$

23. Graph $f(x) = x^2 - 4x + 7$ (From Example 2, Section 3.4 of text.)

24. Graph $P(x) = 400(10 - x)(2 + x)$ (From Example 1, Section 4.7 of text.)

Find an equation for each curve in Problems 25-32.

25. Directrix $x = 0$; focus at $(5, 0)$

26. Directrix $y = 0$; focus at $(0, -3)$

27. Directrix $x - 3 = 0$; vertex at $(-1, 2)$

28. Directrix $y + 4 = 0$; vertex at $(4, -1)$

29. Vertex $(-2, -3)$; focus at $(-2, 3)$

30. Vertex $(-3, 4)$; focus at $(1, 4)$

31. Vertex $(-3, 2)$ and passing through $(-2, -1)$; axis parallel to the y-axis

32. Vertex $(4, 2)$ and passing through $(-3, -4)$; axis parallel to the x-axis

B *Sketch the graph of the polar-form parabola in Problems 33-38. Check your work by finding a Cartesian equation.*

33. $r = \dfrac{6}{1 + \cos \theta}$ **34.** $r = \dfrac{4}{1 - \sin \theta}$

35. $r = \dfrac{-9}{1 + \sin \theta}$ **36.** $r = \dfrac{-2}{1 - \cos \theta}$

37. $r = \dfrac{8}{2 - 2 \cos \theta}$ **38.** $r = \dfrac{9}{3 + 3 \cos \theta}$

Find a polar equation for a parabola with its focus at the pole and with the property given in Problems 39-42.

39. vertex at the polar-form point $(4, 0)$

40. vertex at the polar-form point $(2, \pi)$

41. directrix at $y = -4$ **42.** directrix at $x = 3$

Find a polar equation for the parabola with the Cartesian equation given in Problems 43-46.

43. $y^2 = 4x$ **44.** $x^2 = -2y$

45. $4x^2 = y - 3$ **46.** $x + 1 = 2(y - 3)^2$

47. Find the point(s) on the parabola $y^2 = 9x$ that is (are) closest to $(2, 0)$.

48. Find the point(s) on the parabola $x^2 = 4cy$ that is (are) closest to the focus.

49. Find the equation for the tangent line and the line perpendicular to the parabola $y^2 = 4x$ at the point $(1, -2)$.

50. Find the equation of the set of all points with distances from $(4, 3)$ that equal their distances from $(0, 3)$.

51. Find the equation of the set of all points with distances from (4, 3) that equal their distances from (−2, 1).

52. Find an equation for a parabola whose focal chord has length 6, if it is known that the parabola has focus (4, −2) and its directrix is parallel to the *y*-axis.

53. A parabolic archway has the dimensions shown in Figure 4.13. Find the equation of the parabolic portion.

54. Beams of light parallel to the axis of the parabolic mirror shown in Figure 4.14 strike the mirror and are reflected. Find the distance from the vertex to the point where the beams concentrate, if the radius at the top of the dish is 4 ft.

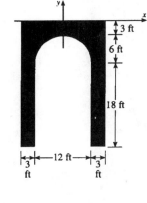

Figure 4.13 A parabolic archway

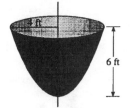

Figure 4.14 A parabolic mirror

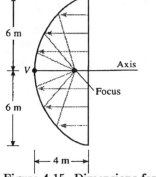

Figure 4.15 Dimensions for a radar antenna

55. A radar antenna is constructed so that a cross section along its axis is a parabola with the receiver at the focus. Find the focus if the antenna is 12 m across and its depth is 4 m. See Figure 4.15.

C 56. Derive the equation of a parabola with $F(0, -c)$, where *c* is a positive number and the directrix is the line $y = c$.

57. Derive the equation of a parabola with $F(c, 0)$, where *c* is a positive number and the directrix is the line $x = -c$.

58. Derive the equation of a parabola with $F(-c, 0)$, where *c* is a positive number and the directrix is the line $x = c$.

59. Show that the length of the focal chord for the parabola $y^2 = 4cx$ is $4c$ $(c > 0)$.

60. Show that the polar equations

$$r = \frac{p}{1 - \cos \theta} \qquad \text{and} \qquad r = \frac{-p}{1 + \cos \theta}$$

represent the same graph.

61. Show that the vertex is the point on a parabola that is closest to the focus.

62. Show that the tangents to a parabola at the two ends of the focal chord intersect on the directrix.

63. Find the area of the triangle formed by the focal chord and the two tangent lines at the end of the focal chord of the parabola $x^2 = 4cy$.

64. Suppose a circle intersects the parabola $x^2 = 4cy$ in four distinct points (x_1, y_1), (x_2, y_2), (x_3, y_3), and (x_4, y_4). Show that

$$x_1 + x_2 + x_3 + x_4 = 0.$$

65. Reflection property of the parabola   Assume that a parabola is given by the equation

$$y = \frac{x^2}{4c}$$

Use Figure 4.16 to prove the reflection property of the parabola by carrying out the indicated steps.

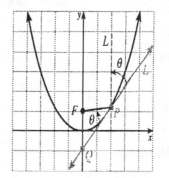

Figure 4.16 Reflection property

a. Find an equation for the tangent line T to the parabola at $P(x_0, y_0)$.

b. Find the coordinates of the point Q where T crosses the y-axis.

c. If F is the focus of the parabola, show that $|\overline{FP}| = |\overline{FQ}|$. Conclude that $\triangle QFP$ is isosceles.

d. Let L be a line parallel to the y-axis. Show that the angle between L and T equals $\theta = \angle FPQ$.

CHAPTER 5

Conic Sections: The Ellipse and the Hyperbola

5.1

ELLIPSES

In this Chapter, we consider two more conic sections, the *ellipse* and the *hyperbola*.

ELLIPSE

> An **ellipse** is the set of all points in the plane the sum of whose distances from two fixed points is a constant.

The fixed points are called the **foci** (plural of **focus**). To see what an ellipse looks like, we will use the special type of graph paper shown in Figure 5.1**a**, where F_1 and F_2 are the foci.

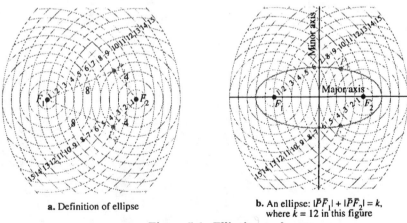

a. Definition of ellipse

b. An ellipse: $|\bar{P}\bar{F}_1| + |\bar{P}\bar{F}_2| = k$, where $k = 12$ in this figure

Figure 5.1 Elliptic graph paper

Let the constant distance be 12. Plot all the points in the plane so that the sum of their distances from the foci is 12. If a point is 8 units from F_1, for example, then it is 4 units from F_2, and you can plot the points P_1 and P_2. The completed graph of this ellipse is shown in Figure 5.1b.

The line passing through F_1 and F_2 is called the **major axis.** The **center** is the midpoint of the segment $\overline{F_1 F_2}$. The **semimajor axis** is the distance from the center to a point of intersection of the ellipse with its major axis. The line passing through the center perpendicular to the major axis is called the **minor axis.** The **semiminor axis** is the distance from the center to a point of intersection of the minor axis with the ellipse. The ellipse is symmetric with respect to the major and minor axes. The intercepts on the major axis are called the **vertices** of the ellipse.

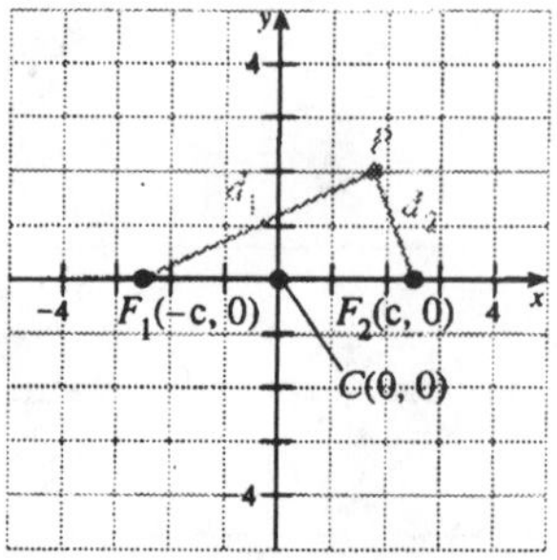

To find the equation of an ellipse, first consider a special case where the center is at the origin. Let the distance from the center to a focus be the positive number c; that is, let $F_1(-c, 0)$ and $F_2(c, 0)$ be the foci and let the constant sum of distances be $2a$ (that is, $d_1 + d_2 = 2a$), as shown in Figure 5.1.

If $P(x, y)$ is any point on the ellipse, then by definition,

$$\left|\overline{PF_1}\right| + \left|\overline{PF_2}\right| = 2a$$

$$\sqrt{(x + c)^2 + (y - 0)^2} + \sqrt{(x - c)^2 + (y - 0)^2} = 2a$$

Simplifying (the details are not shown), we obtain

$$\frac{x^2}{a^2} + \frac{y^2}{a^2 - c^2} = 1$$

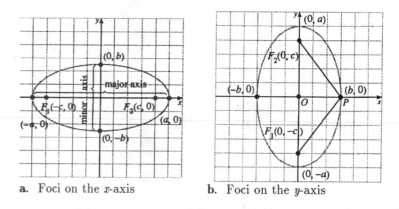

Let $b^2 = a^2 - c^2$ to obtain $\frac{x^2}{a^2} + \frac{y^2}{b^2} = 1$. The graph of this equation is shown in Figure 5.2. Notice that foci are on the major axis, and the intercepts on the minor axis are $(0, b)$ and $(0, -b)$. Also note that $a > c$ and $a > b$.

A similar derivation applies to the ellipse in standard position with foci on the y-axis.

Figure 5.1 Developing the equation of an ellipse by using the definition

a. Foci on the x-axis   **b.** Foci on the y-axis

Figure 5.2 Standard-form ellipses

STANDARD-FORM
EQUATIONS FOR
ELLIPSES

Orientation	Equation	Foci	Constant	Center
Horizontal	$\dfrac{x^2}{a^2} + \dfrac{y^2}{b^2} = 1$	$(-c, 0), (c, 0)$	$2a$	$(0, 0)$
Vertical	$\dfrac{y^2}{a^2} + \dfrac{x^2}{b^2} = 1$	$(0, c), (0, -c)$	$2a$	$(0, 0)$

where $b^2 = a^2 - c^2$ or $c^2 = a^2 - b^2$ with $a > b > 0$

In order to sketch an ellipse, plot the center, the intercepts $\pm a$ on the major axis, and $\pm b$ on the minor axis. Write the equation in standard form, so that there is a 1 on the right and the numerator coefficients of the square terms are also 1. The center is $(0, 0)$; plot the intercepts on the x- and y-axes. For the x-intercepts, plot $\pm$ the square root of the number under the x^2-term; for the y-intercepts, plot $\pm$ the square root of the number under the y^2-term. Finally, draw the ellipse using these intercepts. The longer axis is called the major axis; if this larger axis is horizontal, then the ellipse is horizontal, and if the major axis is vertical, the ellipse is called vertical.

EXAMPLE 1

Graphing an ellipse centered at the origin

Sketch $9x^2 + 4y^2 = 36$. Find the foci.

Solution

First, rewrite the equation in standard form by dividing both sides by 36:
$$\frac{x^2}{4} + \frac{y^2}{9} = 1$$

Because $a^2 = 9$, $b^2 = 4$, the foci are found by calculating c: $c^2 = a^2 - b^2 = 9 - 4 = 5$. Thus, the foci are $(0, -\sqrt{5})$, $(0, \sqrt{5})$. The graph is shown in Figure 5.3. □

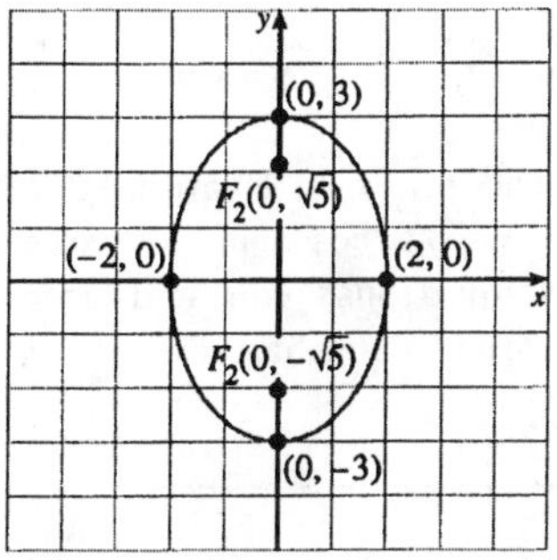

Figure 5.3 Graph of $9x^2 + 4y^2 = 36$

EXAMPLE 2

Parametrize the equation of an ellipse

Graph $2x^2 + 5y^2 = 10$ by using a parametrization.

Solution

One way to parametrize an ellipse is to recall the identity $\cos^2\theta + \sin^2\theta = 1$. We begin by dividing both sides of the given equation by 10:

$$2x^2 + 5y^2 = 10$$

$$\frac{2x^2}{10} + \frac{5y^2}{10} = 1$$

$$\frac{x^2}{5} + \frac{y^2}{2} = 1$$

$$\left(\frac{x}{\sqrt{5}}\right)^2 + \left(\frac{y}{\sqrt{2}}\right)^2 = 1$$

We know that if $\cos\theta = \dfrac{x}{\sqrt{5}}$ and $\sin\theta = \dfrac{y}{\sqrt{2}}$, then $\cos^2\theta + \sin^2\theta = 1$. There are, of course, other choices we could make, but this observation leads us to let $x = \sqrt{5}\cos\theta$ and $y = \sqrt{2}\sin\theta$. You can set up a table of values or use a calculator to obtain the graph shown in Figure 5.4. If you use a table of values, you need only consider values of θ between 0 and $\frac{\pi}{2}$, because we know the ellipse is symmetric with respect to both the major and minor axes.

θ	x	y
0°	2.24	0
15°	2.16	0.37
30°	1.94	0.71
45°	1.58	1
60°	1.12	1.22
75°	0.58	1.37
90°	0	1.41

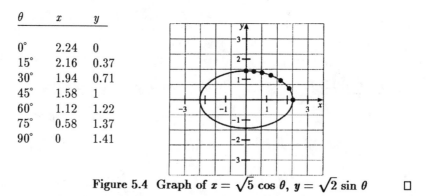

Figure 5.4 Graph of $x = \sqrt{5}\cos\theta,\ y = \sqrt{2}\sin\theta$ □

The parametrization we obtained for the ellipse in Example 2 is not unique. For example, $x = \sqrt{5}\sin\theta$, $y = \sqrt{2}\cos\theta$ is the same ellipse, but if you sketch these parametric equations, you will note that even though the ellipse is the same, the orientation has reversed.

EXAMPLE 3   **Finding the equation of a given ellipse**

Find an equation for the ellipse with foci $(-1, 0)$ and $(1, 0)$ and vertices $(-2, 0)$ and $(2, 0)$.

Solution   By inspection, the center of the ellipse is $(0, 0)$ and the distance to a vertex is 2, so $a = 2$; the distance to a focus is 1, so $c = 1$. We find $b^2 = a^2 - c^2 = 3$. An equation is

$$\frac{x^2}{4} + \frac{y^2}{3} = 1 \qquad\qquad □$$

If an ellipse is not in standard position, but its axes are parallel to the coordinate axes, complete the square to determine the translation. Here is an Example of this procedure.

<u>**EXAMPLE 4**</u>

Solution

Graphing an ellipse by competing the square

Sketch the graph of the equation $9x^2 + 4y^2 - 18x + 16y - 11 = 0$.

Complete the square in both x and y:

$$9x^2 + 4y^2 - 18x + 16y - 11 = 0$$

$$9(x^2 - 2x) + 4(y^2 + 4y) = 11$$

$$9(x^2 - 2x + 1^2) + 4(y^2 + 4y \boxed{+ 2^2}) = 11 + \mathbf{9 \cdot 1} \boxed{+ 4 \cdot 4}$$

$$9(x - 1)^2 + 4(y + 2)^2 = 36$$

$$\frac{(x - 1)^2}{4} + \frac{(y + 2)^2}{9} = 1$$

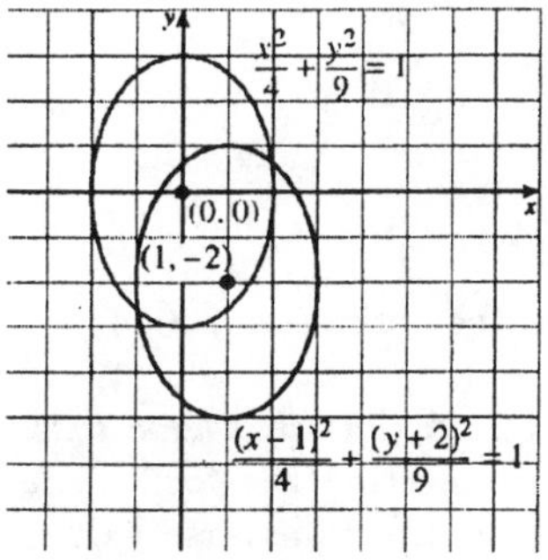

Thus, the graph may be obtained by translating the graph of the ellipse $\frac{x^2}{4} + \frac{y^2}{9} = 1$ by 1 unit to the right and 2 units down. This process is shown in Figure 5.5. First, plot the center $(h, k) = (1, -2)$ and then count out from that point a distance of $a = \pm 3$ (the vertices) on the major axis and label those vertices. Finally, count out from the center the distance $b = \pm 2$ on the minor axis. Using those four points on the ellipse, you can sketch the sketch. ☐

Figure 5.5 Graph of

$9x^2 + 4y^2 - 18x + 16y - 11 = 0$

5.2

HYPERBOLAS

The last of the conic sections to be considered has a definition similar to that of the ellipse.

HYPERBOLA

> A **hyperbola** is the set of all points in the plane such that, for each point on the hyperbola, the difference of its distances from two fixed points is a constant.

The fixed points are called the **foci.** A hyperbola with foci at F_1 and F_2, where the given constant distance is 8 is shown in Figure 5.6.

Art correction; conjugate axis

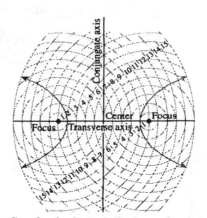

Figure 5.6 Graph of a hyperbola from the definition

The line passing through the foci is called the **transverse axis**. The **center** is the midpoint of the segment connecting the foci. The line passing through the center perpendicular to the transverse axis is called the **conjugate axis**. The transverse axis intersects the hyperbola at points called the **vertices**, and the conjugate axis does not intersect the hyperbola. The hyperbola is symmetric with respect to both the transverse and conjugate axes.

If you use the definition, you can derive the equation for a hyperbola with foci at $(-c, 0)$ and $(c, 0)$ and constant difference $2a$ (both c and a are positive). If (x, y) is any point on the curve, then

$$\left| \sqrt{(x+c)^2 + (y-0)^2} - \sqrt{(x-c)^2 + (y-0)^2} \right| = 2a$$

The procedure for simplifying this expression is left as a problem. After several steps you should obtain

$$\frac{x^2}{a^2} - \frac{y^2}{c^2 - a^2} = 1$$

If $b^2 = c^2 - a^2$, then

$$\frac{x^2}{a^2} - \frac{y^2}{b^2} = 1$$

which is the standard-form equation. Notice that $c^2 = a^2 - b^2$ for the ellipse and that $c^2 = a^2 + b^2$ for the hyperbola. For the ellipse it is necessary that $a^2 > b^2$, but for the hyperbola there is no restriction on the relative sizes for a and b (but c is still less than a for the hyperbola).

Repeat the argument for a hyperbola with foci $F_1(0, c)$ and $F_2(0, -c)$, and you will obtain the other standard-form equation for a hyperbola with a vertical transverse axis.

STANDARD-FORM
EQUATIONS FOR
HYPERBOLAS

	Orientation	Foci	Constant	Center
Horizontal:	$\dfrac{x^2}{a^2} - \dfrac{y^2}{b^2} = 1$	$(-c, 0), (c, 0)$	$2a$	$(0, 0)$
Vertical:	$\dfrac{y^2}{a^2} - \dfrac{x^2}{b^2} = 1$	$(0, c), (0, -c)$	$2a$	$(0, 0)$

where $b^2 = c^2 - a^2$ or $c^2 = a^2 + b^2$

As with the other conics, we shall sketch a hyperbola by determining some information about the curve directly from the equation by inspection. The vertices are located $\pm a$ units from the center. The number $2a$ is the **length of the transverse axis.** The hyperbola does not intersect the conjugate axis, but if you plot the points located $\pm b$ units from the center, you determine a segment on the conjugate axis with length $2b$ called the **length of the conjugate axis.** The endpoints of this segment are useful in determining the shape of the hyperbola.

EXAMPLE 1

Sketching a hyperbola in standard form

Sketch $\dfrac{x^2}{4} - \dfrac{y^2}{9} = 1$.

Solution

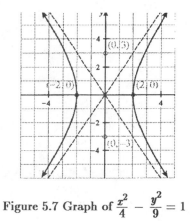

The center of the hyperbola is $(0, 0)$; $a = 2$ and $b = 3$. Plot the vertices at $x = \pm 2$, as shown in the margin. From the form we see that the transverse axis is along the x-axis and the conjugate axis is along the y-axis. Plot the length of the conjugate axis by plotting ± 3 units from the origin. We call these points the **pseudovertices,** because the curve does not actually pass through these points.

Next, form a rectangle by drawing lines through the vertices and pseudovertices parallel to the axes of the hyperbola. This rectangle is called the **central rectangle.** The diagonal lines passing through the corners of the central rectangle are **oblique asymptotes** for the hyperbola, as shown in Figure 5.7; they aid in sketching the hyperbola.

Figure 5.7 Graph of $\dfrac{x^2}{4} - \dfrac{y^2}{9} = 1$ □

For the general hyperbola given by the equation

$$\frac{x^2}{a^2} - \frac{y^2}{b^2} = 1$$

the equations of the oblique asymptotes described are found by replacing the constant term 1 by 0 and then factoring and solving:

$$y = \frac{b}{a} x \qquad \text{and} \qquad y = -\frac{b}{a} x$$

To justify this result, you are asked in Problem 62 to show that the branches of the hyperbola approach $y = \pm \frac{b}{a} x$ as $|x| \to \infty$.

EXAMPLE 2

Completing the square to sketch a hyperbola

Sketch $16x^2 - 9y^2 - 128x - 18y + 103 = 0$.

Solution

Complete the square in both x and y.

$$16x^2 - 9y^2 - 128x - 18y + 103 = 0$$

$$16(x^2 - 8x \quad) - 9(y^2 + 2y \quad) = -103$$

♊ Watch the signs on the second parentheses. ♊

$$16(x^2 - 8x + 4^2) - 9(y^2 + 2y + 1^2) = -103 + 16 \cdot 4^2 - 9 \cdot 1^2$$

$$16(x - 4)^2 - 9(y + 1)^2 = 144$$

$$\frac{(x - 4)^2}{9} - \frac{(y + 1)^2}{16} = 1$$

The graph is shown in Figure 5.8.

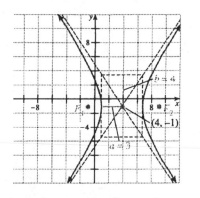

Figure 5.8 Sketch of $16x^2 - 9y^2 - 128x - 18y + 103 = 0$ □

We conclude our discussion of hyperbolas by considering an example in which information about the graph is given and we are asked to find the equation of the hyperbola.

EXAMPLE 3

Equation of a hyperbola given information about the graph

Find the set of points such that, for any point, the difference of its distances from $(6, 2)$ and $(6, -5)$ is always 3.

Solution   From the definition, we see this is a hyperbola with center $(6, -\frac{3}{2})$ and $c = \frac{7}{2}$. Also, $2a = 3$, so $a = \frac{3}{2}$. Because $c^2 = a^2 + b^2$, we have

$$\frac{49}{4} = \frac{9}{4} + b^2 \qquad \text{so that} \qquad b^2 = 10$$

The desired equation is

$$\frac{(y + \frac{3}{2})^2}{\frac{9}{4}} - \frac{(x - 6)^2}{10} = 1 \quad \text{or} \quad \frac{4(y + \frac{3}{2})^2}{9} - \frac{(x - 6)^2}{10} = 1 \qquad \Box$$

5.3

ECCENTRICITY AND POLAR COORDINATES

We defined the parabola as the set of all points P equidistant from a given point F (the focus) and a given line L (the directrix). In other word, for a parabola

$$\frac{\text{DISTANCE FROM P TO F}}{\text{DISTANCE FROM P TO L}} = 1$$

This form of the definition of a parabola is part of the following characterization of conic sections.

ECCENTRICITY

Let F be a point in the plane and let L be a line in the same plane. Then the set of all points P in the plane that satisfy

$$\frac{\text{DISTANCE FROM P TO F}}{\text{DISTANCE FROM P TO L}} = \epsilon$$

is a conic section, and ϵ is a fixed number for each conic called the **eccentricity** of the conic. The conic is

An *ellipse* if $\epsilon < 1$;
A *parabola* if $\epsilon = 1$;
A *hyperbola* if $\epsilon > 1$.

This criteria for a conic are illustrated in Figure 5.9.

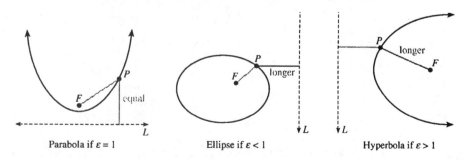

Parabola if $\varepsilon = 1$ Ellipse if $\varepsilon < 1$ Hyperbola if $\varepsilon > 1$

Figure 5.9 Eccentricity characterization of a conic section

Next, we shall examine polar characterizations for the ellipse and the hyperbola that involve the eccentricity ϵ. Consider an ellipse with one focus F at the origin of a polar coordinate plane. Assume that the corresponding directrix L is the vertical line $x = p$, or in polar-form, $\cos \theta = p$ $(p > 0)$ and that the ellipse has eccentricity ϵ. Then, if $P(r, \theta)$ is a polar-form point on the ellipse, we have

$$\epsilon = \frac{\text{DISTANCE FROM P TO F}}{\text{DISTANCE FROM P TO L}} = \frac{r}{p - r \cos \theta}$$

This relationship is shown in Figure 5.10.

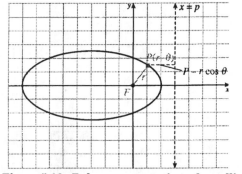

Figure 5.10 Polar representation of an ellipse

Solving for r, we find the ellipse has the polar equation

$$r = \frac{\epsilon p}{1 + \epsilon \cos \theta} \qquad \textit{Because } 0 \leq \epsilon < 1,\ \epsilon \cos \theta \neq -1.$$

Similarly, if the directrix is $x = -p$, the equation is $r = \dfrac{\epsilon p}{1 - \epsilon \cos \theta}$ and if the directrix is $y = p$ or $y = -p$, the corresponding equations are, respectively,

$$r = \frac{\epsilon p}{1 + \epsilon \sin \theta} \qquad \text{and} \qquad r = \frac{\epsilon p}{1 - \epsilon \sin \theta}$$

These four possibilities are summarized in Figure 5.11.

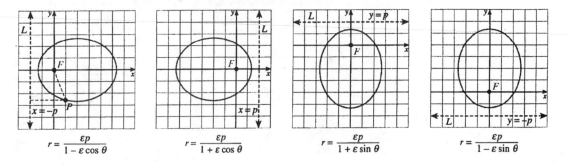

$$r = \frac{\epsilon p}{1 - \epsilon \cos \theta} \qquad\qquad r = \frac{\epsilon p}{1 + \epsilon \cos \theta} \qquad\qquad r = \frac{\epsilon p}{1 + \epsilon \sin \theta} \qquad\qquad r = \frac{\epsilon p}{1 - \epsilon \sin \theta}$$

Figure 5.11 Forms for the equation of an ellipse ($\epsilon < 1$) in standard polar form

EXAMPLE 1	**Describing the graph of an equation in polar form**

Discuss the graph of the polar-form equation $r = \dfrac{2}{2 - \cos\theta}$.

Solution

Begin by writing the equation in standard form: $r = \dfrac{1}{1 - \frac{1}{2}\cos\theta}$.

This form involves a cosine and has $\epsilon = \frac{1}{2} < 1$, so by comparing with the forms in Figure 5.11, we see that the graph must be a horizontal ellipse. The form also tells us that $\epsilon p = 1$, so $p = 2$ and the directrix is $x = -2$. The focus F_1 closer to the directrix is at the pole, and the vertices occur where $\theta = 0$ and $\theta = \pi$. For $\theta = 0$, we obtain $r = 2$, and for $\theta = \pi$, $r = \frac{2}{3}$, so the vertices are the polar points $V_1(\frac{2}{3}, \pi)$ and $V_2(2, 0)$. Since the focus $F_1(0, 0)$ is $\frac{2}{3}$ units to the left of $V_1(2, 0)$, F_2 is the polar point $(\frac{4}{3}, 0)$. The center of the ellipse is midway between the foci, at $(\frac{2}{3}, 0)$, so the minor axis is the vertical line passing through this point. This is the line $x = \frac{2}{3}$ or, in polar form,

$$r\cos\theta = \frac{2}{3} \text{ or } r = \frac{\frac{2}{3}}{\cos\theta}$$

To find the endpoints of the minor axis, we need to solve simultaneously the equations for the axis with the equation of the ellipse; namely,

$$\frac{2}{2 - \cos\theta} = \frac{\frac{2}{3}}{\cos\theta}$$

$$2\cos\theta = \tfrac{2}{3}(2 - \cos\theta)$$

$$\cos\theta = \tfrac{1}{2}$$

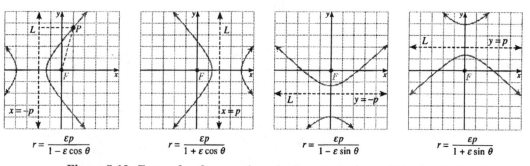

Figure 5.12 Graph of

$$r = \frac{2}{2 - \cos\theta}$$

so that $\theta = \frac{\pi}{3}$ and $\frac{5\pi}{3}$. Solving for r we find $r = \frac{4}{3}$, which gives the vertices $(\frac{4}{3}, \frac{\pi}{3})$ and $(\frac{4}{3}, \frac{5\pi}{3})$. The graph is shown in Figure 5.12. ☐

Formulas for hyperbolas in polar coordinates are obtained in essentially the same way as polar formulas for ellipses. The four different cases that can occur for hyperbolas in standard form are summarized in Figure 5.13.

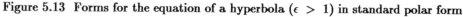

$$r = \frac{\epsilon p}{1 - \epsilon\cos\theta} \qquad r = \frac{\epsilon p}{1 + \epsilon\cos\theta} \qquad r = \frac{\epsilon p}{1 - \epsilon\sin\theta} \qquad r = \frac{\epsilon p}{1 + \epsilon\sin\theta}$$

Figure 5.13 Forms for the equation of a hyperbola ($\epsilon > 1$) in standard polar form

<table>
<tr><td>**EXAMPLE 2**</td><td>**Describing the graph of a hyperbola in polar form**</td></tr>
</table>

Discuss the graph of the polar equation $r = \dfrac{5}{3 + 4\sin\theta}$.

Solution

The standard form of the equation is $r = \dfrac{\frac{5}{3}}{1 + \frac{4}{3}\sin\theta}$. The form tells

us that the eccentricity is $\epsilon = \frac{4}{3}$, and because $\epsilon > 1$, the graph is a hyperbola. We also see that the transverse axis is the y-axis and that because $\epsilon p = \frac{5}{3}$ we have $p = \frac{5}{4}$. Thus the graph has one focus F_1 at the pole and directrix $y = p = \frac{5}{4}$.

The corresponding vertex occurs when $\theta = \frac{\pi}{2}$:

$$r = \frac{5}{3 + 4\sin\frac{\pi}{2}} = \frac{5}{7}$$

so the polar coordinates of this vertex V_1 are $(\frac{5}{7}, \frac{\pi}{2})$. The opposite vertex occurs where $\theta = \frac{3\pi}{2}$ and

$$r = \frac{5}{3 + 4\sin\frac{3\pi}{2}} = -5$$

so the point is $V_2\left(-5, \frac{3\pi}{2}\right)$. Because the vertex V_1 is located $\frac{5}{7}$ units above $F_1(0, 0)$, we find the other focus F_2, will be is located 5 units above the vertex V_1. Thus, F_2 is the polar point $(\frac{40}{7}, \frac{\pi}{2})$. The graph is shown in Figure 5.14. □

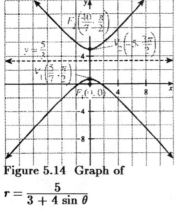

Figure 5.14 Graph of

$r = \dfrac{5}{3 + 4\sin\theta}$

5.4

GEOMETRIC PROPERTIES

Like the parabola, the ellipse has some useful reflection properties. Let P be any point on an ellipse with foci F_1 and F_2, and let T be the tangent line to the ellipse at P, as shown in Figure 5.15.

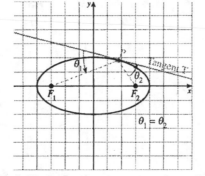

Figure 5.15 Reflection property of an ellipse

Then the line segments $\overline{F_1P}$ and $\overline{F_2P}$ (called the **focal radii**) make equal angles with the tangent line T at P (see Problem 69). The reflection property of an ellipse has the following physical interpretation.

REFLECTIVE PROPERTY OF AN ELLIPSE

> An elliptic mirror has the property that waves emanating from one focus are reflected toward the other focus.

The "whispering room" phenomenon found in many science museums and in famous buildings such as the old U.S. Capitol in Washington, D.C., is an application of this principle. Two people stand at each focus of an elliptic dome. If one person whispers, the other will clearly hear what is said, but anyone *not* near a focus will hear nothing. This is especially impressive if the foci are far apart.

The elliptic reflection principle is also used in a procedure for disintegrating kidney stones. A patient is placed in a tub of water with the shape of an ellipsoid (a three-dimensional elliptic figure) in such a way that the kidney stone is at one focus of the ellipsoid. A pulse generated at the other focus is then concentrated at the kidney stone.

There is also a useful reflection property of hyperbolas. Suppose an aircraft has crashed somewhere in the desert. A device in the wreckage emits a "beep" at regular intervals. Two observers, located at listening posts a known distance apart, time a beep. It turns out that the time difference between the two listening posts multiplied by the velocity of sound gives the value $2a$ for a hyperbola on which the airplane is located. A third listening post will determine two more hyperbolas in a similar fashion, and the airplane must be at the intersection of these hyperbolas. (See Problem 57.)

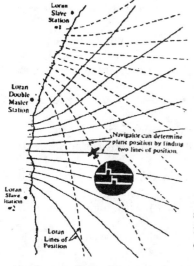

Figure 5.16 LORAN measures the differences in the time of arrival of signals from two sets of stations. The plane's position at the intersection lines is charted on a special map based on a hyperbolic coordinate system.

Diagram courtesy of Bendix Corporation

5.5 PROBLEM SET 5

A   *For each equation in Problems 1-20, sketch each curve.*

1. $\dfrac{x^2}{4} + \dfrac{y^2}{16} = 1$ 2. $\dfrac{x^2}{16} + \dfrac{y^2}{4} = 1$

3. $\dfrac{x^2}{25} - \dfrac{y^2}{9} = 1$ **4.** $\dfrac{x^2}{9} - \dfrac{y^2}{4} = 1$

5. $y^2 - x^2 = 1$ **6.** $x^2 - y^2 = 8$

7. $(x - 1)^2 + 4y^2 = 64$ **8.** $2x^2 + 3y^2 = 12$

9. $x^2 - 2y^2 + 2 = 0$ **10.** $5x^2 - 3y^2 + 15 = 0$

11. $\dfrac{(x - 1)^2}{4} + \dfrac{(y + 3)^2}{16} = 1$ **12.** $\dfrac{(x - 1)^2}{16} + \dfrac{(y + 3)^2}{4} = 1$

13. $4(x - 3)^2 - 9(y + 1)^2 = 36$ **14.** $9(x + 1)^2 - 4(y - 3)^2 = 36$

15. $4x^2 - 9y^2 - 8x + 54y - 41 = 0$ **16.** $x^2 - 4y^2 + 2x + 8y - 7 = 0$

17. $4x^2 + y^2 + 8x - 2y + 4 = 0$ **18.** $x^2 + 4y^2 + 2x - 8y + 4 = 0$

19. $9x^2 + 4y^2 - 8y = 32$ **20.** $4x^2 + 9y^2 - 8x = 32$

Find the standard-form equation for the given information in Problems 21-34.

21. an ellipse with vertices at $(0, 8)$ and $(0, 2)$ and $c = \sqrt{5}$

22. an ellipse centered at the origin with focus at $(0, 3)$; semimajor axis with length 4

23. an ellipse centered at the origin with focus at $(-2, 0)$; minor axis with length 4

24. a hyperbola with foci at $(0, 3)$ and $(0, -3)$ and one vertex at $(0, -2)$

25. a hyperbola with foci at $(\sqrt{2}, 0)$ and $(-\sqrt{2}, 0)$ and one vertex at $(1, 0)$

26. a hyperbola with vertices at $(5, 0)$ and $(-5, 0)$ and one focus at $(-7, 0)$

27. a conic with major axis $-4 \le x \le 4$ and minor axis $-3 \le y \le 3$

28. a conic with transverse axis $-3 \le x \le 3$ and conjugate axis $-4 \le y \le 4$

29. an ellipse with center at $(2, 1)$ semiminor axis with length 3 and vertices at $(2, 6)$ and $(2, -4)$

30. a conic with foci at $(-1, 0)$ and $(1, 0)$ with major axis with length 12

31. a conic with foci at $(0, 6)$ and $(0, -6)$ with transverse axis with length 4

32. the set of points such that for any point, the sum of its distances from $(4, -3)$ and $(-4, -3)$ is 12

33. the set of points such that, for any point, the difference of its distances from $(4, -3)$ and $(-4, -3)$ is 6

34. a hyperbola with vertices $(3, 0)$, $(-3, 0)$ and asymptotes $y = 3x$ and $y = -3x$

B   *Sketch the graph of each polar-form equation in Problems 35-38.*

35. $r = \dfrac{4}{6 + \cos \theta}$ 36. $r = \dfrac{4}{2 - 3 \sin \theta}$

37. $r = \dfrac{5}{1 - 2 \sin \theta}$ 38. $r = \dfrac{-3}{2 - \sin \theta}$

39. Find an equation for the tangent line to the ellipse $5x^2 + 4y^2 = 56$ at the point $(-2, 3)$.

40. Find an equation of an ellipse that is tangent to the coordinate axes and to the line $y = 6$. Is there only one such ellipse?

41. Find two points on the ellipse $\dfrac{x^2}{4} + y^2 = 1$ where the tangent line also passes through the point $(0, -2)$.

42. Find the smallest distance from the point $(2, 0)$ to the ellipse $3x^2 + 2y^2 + 6x - 3 = 0$

43. Find an equation for the hyperbola with vertices $(3, -1)$, $(-1, -1)$, and asymptotes $y = \frac{9}{4}x - \frac{13}{4}$ and $y = -\frac{9}{4}x + \frac{5}{4}$.

44. Find an equation for the hyperbola with vertices $(9, 0)$, $(-9, 0)$ whose asymptotes are perpendicular to each other.

45. Find an equation for a hyperbola in standard position that contains the points $\left(3, \dfrac{\sqrt{5}}{2}\right)$ and $(-2, 0)$.

46. Show that $x = x_0 + a \cosh t$, $y = y_0 + b \sinh t$ are parametric equations for one branch of a hyperbola. Find a Cartesian equation for this hyperbola. *Note:* This is the reason sinh t and cosh t are called "hyperbolic functions." (See Section 7.8 of the text.)

47. Show that the equations $x = x_0 + a \sinh t$ and $y = y_0 + b \cosh t$ are parametric equations for one branch of a hyperbola. (See Section 7.8 of the text.)

48. An **equilateral hyperbola** is one with an equation of the general form $y^2 - x^2 = a^2$. Show that a hyperbola is equilateral if and only if its asymptotes are perpendicular to each other.

49. The orbit of a planet is an ellipse whose major axis and minor axis are, respectively, 100 million and 81 million miles long. Find an equation for the ellipse. How far apart are its foci?

SCALE
Astronomical Units
0 400 800 1200
200 600 1000
Millions of Miles

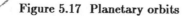

Figure 5.17 Planetary orbits

50. The orbit of the earth around the sun is elliptical with the sun at one focus. The semimajor axis of this orbit is 9.3×10^7 mi and the eccentricity is about 0.017. Determine the greatest and least distance of the earth from the sun (correct to two significant digits). *Hint*: Use polar coordinates.

C 51. Consider a person A who fires a rifle at a distant gong B. Assuming that the ground is flat, where must you stand to hear the sound of the gun and the sound of the gong simultaneously? *Hint*: To answer this question, let x be the distance that sound travels in the length of time it takes the bullet to travel from the gun to the gong. Show that the person who hears the sounds simultaneously must stand on a branch of a hyperbola (the one nearest the target) so that the difference of the distances from A to B is x.

52. Three LORAN stations are located at $(4, 0)$, $(0, 0)$, and $(4, \frac{\pi}{4})$ in a polar coordinate system. Radio signals are sent out from all three stations simultaneously. An airplane receiving the signals notes that the signals from the second and third stations arrive at $\frac{2}{c}$ seconds later than the signal from the first, where c is the velocity of a radio signal. What is the location (in polar coordinates) of the airplane?

53. Derive the equation for the ellipse with foci $F_1(-c, 0)$ and $F_2(c, 0)$ for $c > 0$ and constant distance $2a$.

54. Derive the equation for the hyperbola with foci $F_1(-c, 0)$ and $F_2(c, 0)$ for $c > 0$ and constant distance $2a$.

55. Find conditions on the coefficients of the equation

$$Ax^2 + Cy^2 + Dx + Ey + F = 0$$

with $AC > 0$ that guarantee the graph of the equation will be

a. a line
b. an ellipse
c. a circle
d. a point
e. a hyperbola
f. no graph

56. a. Show that the tangent line to the ellipse

$$\frac{x^2}{a^2} + \frac{y^2}{b^2} = 1$$

at the point (x_0, y_0) has the equation

$$\frac{x_0 x}{a^2} + \frac{y_0 y}{b^2} = 1$$

b. Use part a to show that a tangent line to a vertex of an ellipse in standard form is either vertical or horizontal.

57. Prove that the lines $y = \frac{b}{a} x$ and $y = -\frac{b}{a} x$ are asymptotes of the hyperbola

$$\frac{x^2}{a^2} - \frac{y^2}{b^2} = 1$$

Hint: Show that the vertical distance $D(x)$ between $y = \frac{b}{a}\sqrt{x^2 - a^2}$ and the line $y = \frac{b}{a}x$ tends to 0 as $x \to \infty$, as shown in Figure 5.18.

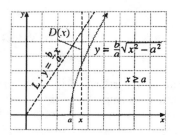

Figure 5.18 Problem 57

CHAPTER 6

Curve Sketching

When sketching a curve, we first check to see whether it is a type of curve we recognize (a line, a conic, a trigonometric function, or something else). If it is not a curve we recognize, we ultimately sketch it by plotting some points; but *before* we plot points we find out as much about the curve as we can. To do this we check (1) symmetry, (2) extent, (3) asymptotes, and (4) intercepts.

6.1

Symmetry

One of the most valuable tools in curve sketching is symmetry. This section deals with recognizing when a curve is symmetric by performing a simple test on its equation. In general, two points are symmetric with respect to a line if that line is the perpendicular bisector of the line containing the two points. In this book, we will check for symmetry with respect to the coordinate axes and the origin (see Figure 6.1).

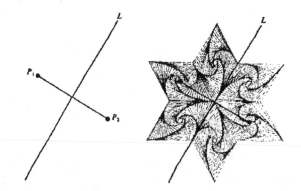

Figure 6.1   (a) Points P_1 and P_2 are symmetric with respect to L.
(b) The star is symmetric with respect to L.

Figure 6.2 shows an example of a curve that is symmetric with respect to the x-axis. Whenever (x, y) is on the curve, so is $(x, -y)$. This gives us a simple algebraic test: *if the equation remains unchanged when y is replaced by $-y$, then it is symmetric with respect to the x-axis*, because (x, y) and $(x, -y)$ must both satisfy the equation.

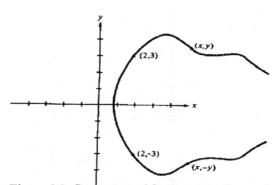

Figure 6.2 Symmetry with respect to the x-axis

Similarly, Figure 6.3 shows a curve that is symmetric with respect to the y-axis. Whenever (x, y) is on the curve, so is $(-x, y)$. This means that, *if the equation is unchanged when x is replaced by $-x$, then it is symmetric with respect to the y-axis.*

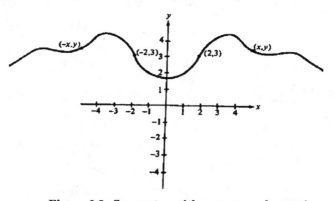

Figure 6.3 Symmetry with respect to the y-axis

Finally, Figure 6.4 shows a curve symmetric with respect to the origin. Whenever (x, y) is on the curve, so is $(-x, -y)$. This means, *if the equation is unchanged when x and y are replaced by $-x$ and $-y$, respectively, then it is symmetric with respect to the origin.*

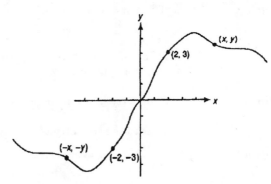

Figure 6.4 Symmetry with respect to the origin

SYMMETRY

> A curve is symmetric with respect to
>
> **the x-axis** if the equation remains unchanged when y is replaced
> by $-y$;
> **the y-axis** if the equation remains unchanged when x is
> replaced by $-x$;
> **the origin** if the equation remains unchanged when x and y are
> simultaneously replaced by $-x$ and $-y$, respectively. Also,
> the curve is symmetric with respect to the origin if
> it is symmetric with respect to both the x- and y-axes.

EXAMPLE 1   **Symmetry of a quadratic function**

Check the symmetry of $x^2 + 2xy^2 + y^2 = 4$.

Solution   This curve is symmetric with respect to the x-axis, since

$$x^2 + 2x(-y)^2 + (-y)^2 = 4$$

is the same as the original equation. It is not symmetric with respect
to the y-axis or the origin.   □

EXAMPLE 2   **Symmetry with a trigonometric function**

Check the symmetry of $y = \cos x$.

Solution   This curve is symmetric with respect the the y-axis, since

$$y = \cos(-x)$$

is the same as the original equation. It is not symmetric with respect
to the x-axis or the origin.   □

EXAMPLE 3   **Symmetry of a third degree equation**

Check the symmetry of $x^3 + 2xy^2 + 4x^2y + 3y^2 = 0$.

Solution   This curve is symmetric with respect to the origin, since

$$(-x)^3 + 2(-x)(-y)^2 + 4(-x)^2(-y) + 3(-y)^2 = 0$$

is the same as the original equation if we multiply both sides by -1.
It is not symmetric with respect to the x- or y-axes.   □

EXAMPLE 4   **Symmetry of a fourth degree equation**

Check the symmetry of $x^2 + 5x^2y^2 = 5$.

Solution   This curve is symmetric with respect to the x-axis, since

$$x^2 + 5x^2(-y)^2 = 5$$

is the same as the original equation. It is also symmetric with respect to the *y*-axis, since

$$(-x)^2 + 5(-x)^2y^2 = 5$$

is the same as the original equation. If a curve is symmetric with respect to both the *x*- and *y*-axes, then it must also be symmetric with respect to the origin.   □

6.2

Extent

By *extent*, we mean the domain and range of a curve. If certain values of one or the other variable cause division by zero or imaginary values, those values must be excluded.

PROCEDURE FOR FINDING THE EXTENT

> **The domain is the set of all possible replacements for *x*. To find the domain:**
>   a. Solve for *y* (if possible).
>   b. The domain is the set of all real values for *x* except those that
>   i.   lead to division by 0; or
>   ii. cause a negative number under a square root (or other even-indexed root).
>
> **The range is the set of all possible replacements for *y*. To find the range:**
>   a. Solve for *x* (if possible).
>   b. The range is the set of all real values for *y* except those that
>   i.   lead to division by 0; or
>   ii. cause a negative number under a square root (or other even-indexed root).

EXAMPLE 1   **Extent of a rational function**

Find the domain and range of $y = \frac{4}{x}$.

Solution   The domain is the set of all real numbers *except* 0, since this value causes division by zero. In such a case, we write all reals, $x \neq 0$. For the range, solve for *x*:

$$x = \frac{4}{y}$$

The range is also the set of all real numbers, $y \neq 0$.   □

EXAMPLE 2	**Extent of a rational function**

Find the domain of $y = \dfrac{(x - 3)(x - 2)(x - 4)}{(x + 2)(x - 4)}$.

Solution

The domain is the set of all real numbers, $x \neq -2$ and $x \neq 4$.   □

EXAMPLE 3	**Extent of a radical function**

Find the domain and range of $y = \sqrt{\dfrac{x}{x - 1}}$.

Solution

For the domain:

$x \neq 1$ *Exclude values that lead to division by 0.*

$\dfrac{x}{x - 1} \geq 0$ *Numbers under a square root must be non-negative.*

x: $-$ | $+$ | $+$
$x - 1$: $-$ | $-$ | $+$ } *Signs of factors*

         0   1

Domain: $(-\infty, 0] \cup (1, \infty)$

For the range, we solve for x:

$$y^2 = \frac{x}{x - 1}$$
$$y^2(x - 1) = x$$
$$y^2 x - y^2 = x$$
$$y^2 x - x = y^2$$
$$(y^2 - 1)x = y^2$$
$$x = \frac{y^2}{y^2 - 1}$$

We must exclude values that lead to division by 0:

$$y^2 - 1 \neq 0$$
$$y \neq 1, -1$$

But y cannot be negative, since y is a square root, which is non-negative. Therefore, the range is the set of all non-negative real numbers except $y = 1$.   □

6.3

Asymptotes

An asymptote is a line such that, as a point P on the curve moves farther away from the origin, the distance between P and the asymptote tends toward 0. In this section, we are concerned with finding horizontal, vertical, and oblique asymptotes. The key to finding horizontal and vertical asymptotes is to find values that cause division by 0.

VERTICAL ASYMPTOTES

For easy reference, here are the examples of Section 6.2.

Example 1: $y = \frac{4}{x}$
Example 2:

$$y = \frac{(x - 3)(x - 2)(x - 4)}{(x + 2)(x - 4)}$$

Example 3: $y = \sqrt{\frac{x}{x - 1}}$

Solve the equation for $y = \frac{P(x)}{D(x)}$, where the fraction is reduced. If r is a value for which $D(r) = 0$, then $x = r$ is a vertical asymptote. In Example 1 of Section 6.2, there is a vertical asymptote at $x = 0$, and in Example 2 of Section 6.2, there are vertical asymptotes at $x = -2$. The value $x = 4$ is a deleted point and not an asymptote. In Example 3, the vertical asymptote has equation $x = 1$.

HORIZONTAL ASYMPTOTES

Solve the equation for $x = \frac{Q(y)}{S(y)}$, if possible. If r is a value for which $S(r) = 0$, then $y = r$ is a horizontal asymptote. In Example 1 of Section 6.2, it is easy to solve for x to find a horizontal asymptote $x = 0$. Sometimes it is not convenient (or possible) to solve an equation for x, as you can see by looking at Examples 2 and 3 of Section 6.2.

Using calculus, it can be shown that, if

$$y = \frac{P(x)}{D(x)}$$

where $P(x)$ and $D(x)$ are polynomial functions of x with no common factors (that is, where the rational expression is reduced), the asymptotes depend on the degree of P and D. Suppose that $P(x)$ has degree M with a leading coefficient p, and that $D(x)$ has degree N with leading coefficient d. The asymptotes can then be found according to the rules given in the following box.

OBLIQUE ASYMPTOTES

Notice that we included oblique (slant) asymptotes in the above box. A curve will have a oblique asymptote whenever the degree of the numerator is one more than the degree of the denominator. Once again, consider Example 2 of Section 6.2,

$$y = \frac{x^2 - 5x + 6}{x + 2}, \, x \neq 4$$

Notice that $M > N$ so there is no horizontal asymptote, but $M = 2$ and $N = 1$, so the degree of the numerator is one more than the degree of the denominator. Divide to find

$$y = x - 7 + \frac{20}{x + 2}$$

Disregard the remainder term to find the oblique asymptote, namely $y = x - 7$.

PROCEDURE FOR FINDING ASYMPTOTES

Vertical asymptotes: $x = r$, where r is a value that causes division by 0 when the equation is solved for y and is reduced.

Horizontal asymptotes: $y = r$, where r is a value that causes division by 0 when the equation is solved for x and is reduced. Sometimes it is not possible (or convenient) to solve for x. If the equation is solved for y and is reduced, then

$y = 0$ is a horizontal asymptote if $M < N$

$y = \frac{p}{d}$ is a horizontal asymptote if $M = N$

no horizontal asymptote exists if $M > N$

oblique asymptotes: Solve the equation for y and reduce; then

$y = mx + b$ is a oblique asymptote if $M = N + 1$

where $mx + b$ is the quotient (disregard the remainder) obtained when $P(x)$ is divided by $D(x)$.

EXAMPLE 1

Asymptotes for a rational function

Find the vertical, horizontal, and oblique asymptotes for

$$y = \frac{6x^2 - x - 1}{4x^2 - 4x + 1}$$

Solution   First, make sure the rational function is reduced:

$$y = \frac{6x^2 - x - 1}{4x^2 - 4x + 1}$$
$$= \frac{(3x + 1)(2x - 1)}{(2x - 1)^2}$$
$$= \frac{3x + 1}{2x - 1}$$

Vertical asymptote: can be found when $2x - 1 = 0$; $x = \frac{1}{2}$ is the equation of a vertical asymptote.
Horizontal asymptote: $y = \frac{3}{2}$, which is found by looking at the leading coefficients of the reduced form.
Oblique asymptotes: do not exist for this curve, since the degree of the numerator is not one more than the degree of the denominator. ☐

EXAMPLE 2

Asymptotes for a rational function

Find the vertical, horizontal, and oblique asymptotes for

$$y = \frac{3x^3 - 2x^2 - 4x + 6}{x^2 - 3}$$

Solution   If you use long division,

$$
\begin{array}{r}
3x - 2 \\
x^2 - 3 \,) \overline{\smash{)}\, 3x^3 - 2x^2 - 4x + 6} \\
\underline{3x^3 \qquad\quad - 9x} \\
-2x^2 + 5x + 6 \\
\underline{-2x^2 \qquad + 6} \\
5x
\end{array}
$$

Since there is a remainder, the rational expression is reduced.

Vertical asymptotes: can be found when $x^2 - 3 = 0$; $x = \pm\sqrt{3}$.

Horizontal asymptotes: none, since the degree of the numerator is larger than the degree of the denominator.

Oblique asymptotes: exist when the degree of the numerator is one more than the degree of the denominator. Carry out the long division (as shown above) and disregard the remainder; the oblique asymptote is

$$y = 3x - 2 \qquad\qquad \square$$

It is possible that a curve cross its oblique asymptote. For Example 2, you could find the point of intersection by solving

$$3x - 2 = \frac{3x^3 - 2x^2 - 4x + 6}{x^2 - 3}$$

$$(x^2 - 3)(3x - 2) = 3x^3 - 2x^2 - 4x + 6$$

$$0 = 5x$$

$$0 = x$$

The point where the curve crosses its asymptote is $(0, -2)$.

6.4

Intercepts

The intercepts are the places where the curve crosses the coordinate axes. When finding the intercepts, we are really plotting points on the curve, but these are generally the easiest points to find.

PROCEDURE FOR
FINDING INTERCEPTS

> **x-intercepts:** set $y = 0$ and solve for x (if possible)
> **y-intercepts:** set $x = 0$ and solve for y (if possible)

EXAMPLE 1   **Curve sketching**

Sketch the curve $x^2(y - 2) = 2$, given the following information:

> Symmetry: with respect to the y-axis
> Domain: $x \neq 0$
> Range: $y > 2$
> Asymptotes: $x = 0$, $y = 2$
> Intercepts: none

Solution   Since we have symmetry with respect to the y-axis, we focus our attention on Quadrants I and II, since the symmetry will give us the curve in Quadrants III and IV. We draw the asymptotes and limit our attention to the domain and range. It is customary to darken the parts of the plane **not** included in the domain or range, as shown by the shaded portion of Figure 6.5(a). Since there are no intercepts, we plot the following points:

$$x = 1, \text{ then } 1(y - 2) = 2, \text{ so } y = 4$$
$$x = 2, \text{ then } 4(y - 2) = 2, \text{ so } y = \frac{5}{2}$$

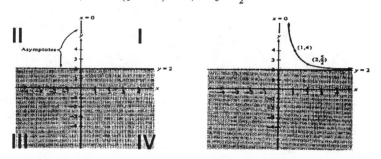

Figure 6.5 Preliminary sketch

We plot these points and, using the fact that $x = 0$ and $y = 2$ are asymptotes, sketch the part of the curve in the first quadrant, as shown in Figure 6.5(b). By symmetry, we then sketch the rest of the curve, as shown in Figure 6.6.

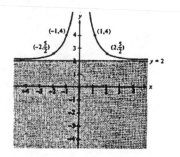

Figure 6.6 Graph of $x^2(y - 2) = 2$ □

EXAMPLE 2   **Sketching a rational function**

Sketch $x^2 = \dfrac{1 + y^2}{1 - y^2}$.

Solution   **Symmetry:**   The curve is symmetric with respect to the x-axis, since

$$x^2 = \frac{1 + (-y)^2}{1 - (-y)^2}$$

is the same as the original equation; it is also symmetric with respect to the y-axis and the origin.

Extent: **domain:** Solve for y.

$$x^2 = \frac{1 + y^2}{1 - y^2}$$

$$x^2 - x^2 y^2 = 1 + y^2$$

$$x^2 - 1 = y^2 + x^2 y^2$$

$$x^2 - 1 = (1 + x^2) y^2$$

$$y^2 = \frac{x^2 - 1}{x^2 + 1}$$

$$y = \pm \sqrt{\frac{x^2 - 1}{x^2 + 1}}$$

We must now rule out values of x that cause division by 0 or negative values under the square root radical. Solve

$$\frac{x^2 - 1}{x^2 + 1} \geq 0$$

The solution for this inequality gives the domain:

$$x \leq -1 \quad \text{or} \quad x \geq 1$$

range: Solve for x.

$$x = \pm \sqrt{\frac{1 + y^2}{1 - y^2}}$$

We need to find the y-values that cause division by 0 or negative values under the square root radical. First solve

$$1 - y^2 = 0$$

to find $y = \pm 1$. These are *excluded values*. Next solve

$$\frac{1 + y^2}{1 - y^2} \geq 0, \quad y \neq \pm 1$$

to find the range: $-1 < y < 1$

Use the information about extent to darken (shade)
the portions of the plane that *cannot* contain the
graph. This is shown in Figure 6.7.

Asymptotes: **vertical:**

$$y = \pm \frac{\sqrt{x^2 - 1}}{\sqrt{x^2 + 1}}$$ *No division by 0, so there are no vertical asymptotes.*

horizontal:

$$x = \pm \frac{\sqrt{1 + y^2}}{\sqrt{1 - y^2}}$$ *Division by 0 when $y = \pm 1$.*

Horizontal asymptotes are $y = 1$, $y = -1$; draw these as
dashed lines on the coordinate axes.

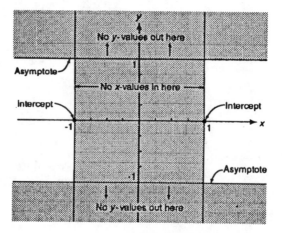

Figure 6.7

Intercepts: ***x*-intercepts:** if $y = 0$, then $x = \pm 1$; so the x-intercepts
are at $(1, 0)$ and $(-1, 0)$.

***y*-intercepts:** if $x = 0$, then $y = \pm \sqrt{-1}$; so there are no
y-intercepts.

Plot points: Plot one or more relevant points, and make use of the
information obtained above to sketch the curve, as
shown in Figure 6.8.

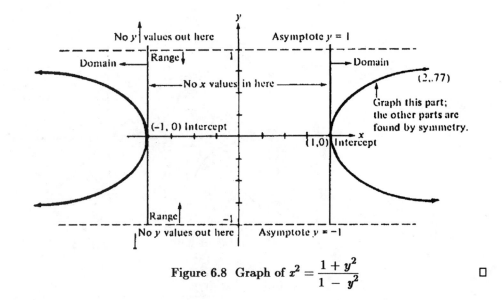

Figure 6.8 Graph of $x^2 = \dfrac{1 + y^2}{1 - y^2}$ □

EXAMPLE 3   **Curve sketching**

Sketch $xy + 4y = 1$.

Solution   **Symmetry:** none

Extent: **domain:** Since $y = \dfrac{1}{x + 4}$, the domain is the set of all
real numbers, except $x = -4$

range: Since $x = \dfrac{1 - 4y}{y}$, the range is the set of all
real numbers, except $y = 0$.

Asymptotes: $x = -4$ and $y = 0$ are the asymptotes.

Intercepts: x-intercept: none (since $y \neq 0$);

y-intercept: $(0, \frac{1}{4})$

Plot points and make use of the information above to sketch the curve, as shown in Figure 6.9.

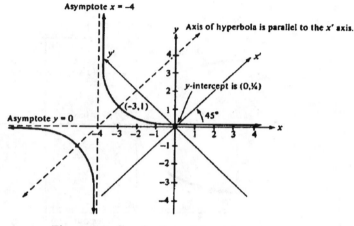

Figure 6.9 Graph of $xy + 4y = 1$ □

6.5 PROBLEM SET 6

A

In Problems 1–4, use the given information and plot some points to draw a sketch of the curve.

1. $x^2 y = 4$
   Symmetry: with respect to the y-axis
   Domain: $x \neq 0$
   Range: $y > 0$
   Asymptotes: $x = 0$; $y = 0$
   Intercepts: none

2. $xy^2 - y^2 - 1 = 0$
   Symmetry: with respect to the x-axis
   Domain: $x > 1$
   Range: $y \neq 0$
   Asymptotes: $x = 1$; $y = 0$
   Intercepts: none

3. $x^2 y - 4x + 2y = 0$
   Symmetry: with respect to the origin
   Domain: all real numbers
   Range: $(-\sqrt{2}, \sqrt{2})$
   Asymptotes: $y = 0$
   Intercept: $(0, 0)$

4. $x^2 y^2 - 4xy^2 + 3y^2 - 4 = 0$
   Symmetry: with respect to the x-axis
   Domain: $(-\infty, 1) \cup (3, \infty)$
   Range: $y \neq 0$
   Asymptotes: $x = 1$; $x = 3$; $y = 0$
   Intercepts: $(0, \frac{2}{3}\sqrt{3})$, $(0, -\frac{2}{3}\sqrt{3})$

Find the symmetry, extent, asymptotes, and intercepts for the curves whose equations are given in problems 5–20.

5. $xy = 2$

6. $xy = 6$

7. $y = \dfrac{x + 1}{x}$

8. $y = \dfrac{x + 1}{x + 2}$

9. $y = \dfrac{2x^2 + x - 10}{x + 2}$

10. $y = \dfrac{3x^2 + 5x - 2}{x + 2}$

11. $y = \dfrac{2x^3 - 3x^2 - 2x}{2x + 1}$

12. $y = \dfrac{x^3 + 6x^2 + 15x + 14}{x + 2}$

13. $9x^2 + 4y^2 - 36 = 0$

14. $6x^2 - 2y^2 + 10 = 0$

15. $13x^2 - 10xy + 13y^2 - 72 = 0$ **16.** $y^2 x - 2y^2 + 2 = 0$

17. $x^2 y - 4xy + 3y - 4 = 0$ **18.** $x^3 - y^2 - 4y = 0$

19. $x^4 - x^2 y^2 - 4x^2 + y^2 = 0$ **20.** $2y^2 - xy^2 + x - 1 = 0$

Graph the curves in Problems 21–36. Note that these curves are the same curves that are given in Problems 5-20.

21. $xy = 2$

22. $xy = 6$

23. $y = \dfrac{x + 1}{x}$

24. $y = \dfrac{x + 1}{x + 2}$

25. $y = \dfrac{2x^2 + x - 10}{x + 2}$

26. $y = \dfrac{3x^2 + 5x - 2}{x + 2}$

27. $y = \dfrac{2x^3 - 3x^2 - 2x}{2x + 1}$

28. $y = \dfrac{x^3 + 6x^2 + 15x + 14}{x + 2}$

29. $9x^2 + 4y^2 - 36 = 0$

30. $6x^2 - 2y^2 + 10 = 0$

31. $13x^2 - 10xy + 13y^2 - 72 = 0$ **32.** $y^2 x - 2y^2 + 2 = 0$

33. $x^2 y - 4xy + 3y - 4 = 0$ **34.** $x^3 - y^2 - 4y = 0$

35. $x^4 - x^2 y^2 - 4x^2 + y^2 = 0$ **36.** $2y^2 - xy^2 + x - 1 = 0$

37. Graph $f(x) = 1 + 2x + \dfrac{18}{x}$

From Problem Set 4.4,
Problem 9 of text.

38. Graph $f(x) = \dfrac{1}{x^2 + 3}$

From Problem Set 4.4,
Problem 18 of text.

39. Graph $y = \dfrac{x^2 - x - 2}{x - 3}$

From Example 8, Section
4.5 of text.

40. Graph $y = \dfrac{x^3 + 1}{x^3 - 8}$

From Problem Set 4.5,
Problem 29 of text.

CHAPTER 7

Catalog of Special Curves

Boldface curves are fundamental, and are listed on Table 1.3 of text. For some of these curves we give the area enclosed (A) and the arc length s. Rectangular curves can be translated to (h, k) by replacing x by $x' - h$ and y by $y' - k$. Polar-form curves can be rotated through an angle α by replacing $\theta = \theta' - \alpha$. The particular values for the constants used for each graph are shown in parentheses.

Absolute value function

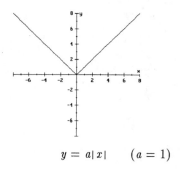

$$y = a|x| \qquad (a = 1)$$

Archimedian spiral (see Spiral of Archimedes)

Arccosecant, arccosine, arccotangent, arcsecant, arcsine, and arctangent functions (see inverse functions)

Astroid

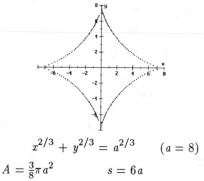

$$x^{2/3} + y^{2/3} = a^{2/3} \qquad (a = 8)$$
$$A = \tfrac{3}{8}\pi a^2 \qquad\qquad s = 6a$$

This curve is described by a point P on a circle of radius $a/4$ as it rolls on the inside of a circle of radius a.

Bifolium

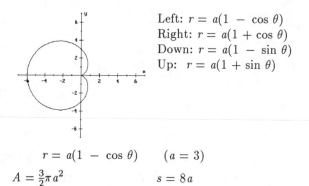

$$(x^2 + y^2)^2 = ax^2y \quad \text{or} \quad r = a\sin\theta\cos^2\theta \qquad (a = 12)$$

Cardioid (this is also a special case of an epicycloid)

Left: $r = a(1 - \cos\theta)$
Right: $r = a(1 + \cos\theta)$
Down: $r = a(1 - \sin\theta)$
Up: $r = a(1 + \sin\theta)$

$$r = a(1 - \cos\theta) \qquad (a = 3)$$
$$A = \tfrac{3}{2}\pi a^2 \qquad\qquad s = 8a$$

This is the curve described by a point P of a circle of radius a as it rolls on the outside of a fixed circle of radius a. This curve is also a special case of a limaçon.

Cassinian curves (see Ovals of Cassini)

Catenary (see hyperbolic cosine)

Circle

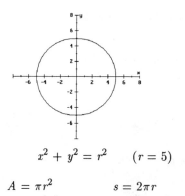

$$x^2 + y^2 = r^2 \qquad (r = 5)$$

$$A = \pi r^2 \qquad s = 2\pi r$$

Cissoid of Diocles

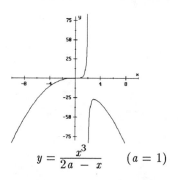

$$y = \frac{x^3}{2a - x} \qquad (a = 1)$$

This is a curve described by a point P such that the $|OP| = |RS|$. It is used in the problem of *duplication of a cube* (*i.e.*; finding the side of a cube which is twice the volume of a given cube).

Cochleoid (or Ouija board curve)

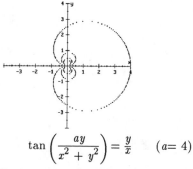

$$\tan\left(\frac{ay}{x^2 + y^2}\right) = \frac{y}{x} \qquad (a = 4)$$

Conic section (see circle, ellipse, hyperbola, and parabola)

Cosecant function

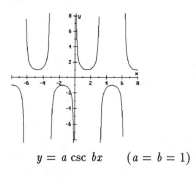

$$y = a \csc bx \qquad (a = b = 1)$$

Cosine function

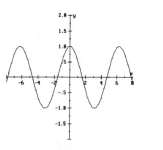

$$y = a \cos bx \qquad (a = b = 1)$$

Cotangent function

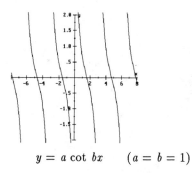

$$y = a \cot bx \qquad (a = b = 1)$$

Cube root function

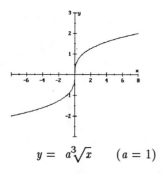

$$y = a\sqrt[3]{x} \qquad (a = 1)$$

Cubical parabola

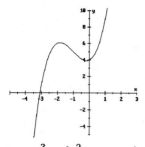

$$y = ax^3 + bx^2 + cx + d$$
$$(a = 1,\ b = 3,\ c = 1,\ d = 4)$$

Cubic function

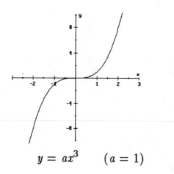

$$y = ax^3 \qquad (a = 1)$$

Curate cycloid (see Trochoid)

Cycloid

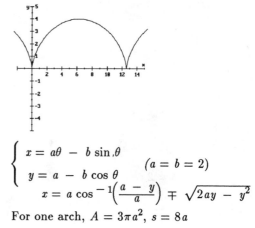

$$\begin{cases} x = a\theta - b\sin\theta \\ y = a - b\cos\theta \end{cases} (a = b = 2)$$
$$x = a\cos^{-1}\left(\frac{a - y}{a}\right) \mp \sqrt{2ay - y^2}$$

For one arch, $A = 3\pi a^2$, $s = 8a$

Deltoid

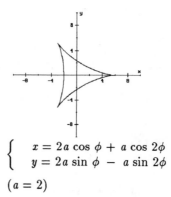

$$\begin{cases} x = 2a\cos\phi + a\cos 2\phi \\ y = 2a\sin\phi - a\sin 2\phi \end{cases}$$

$$(a = 2)$$

Ellipse

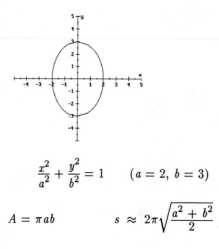

$$\frac{x^2}{a^2} + \frac{y^2}{b^2} = 1 \qquad (a = 2,\ b = 3)$$

$$A = \pi ab \qquad\qquad s \approx 2\pi\sqrt{\frac{a^2 + b^2}{2}}$$

Epicycloid

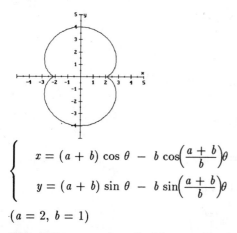

$$\begin{cases} x = (a + b)\cos\theta - b\cos\left(\dfrac{a + b}{b}\right)\theta \\[2mm] y = (a + b)\sin\theta - b\sin\left(\dfrac{a + b}{b}\right)\theta \end{cases}$$

$(a = 2,\ b = 1)$

This is the curve described by a point P on a circle of radius b as it rolls on the outside of a circle of radius a.

Evolute of an ellipse

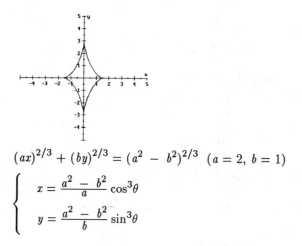

$(ax)^{2/3} + (by)^{2/3} = (a^2 - b^2)^{2/3}$ $(a = 2,\ b = 1)$

$$\begin{cases} x = \dfrac{a^2 - b^2}{a}\cos^3\theta \\[2mm] y = \dfrac{a^2 - b^2}{b}\sin^3\theta \end{cases}$$

Exponential curve

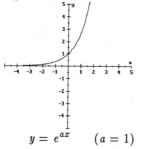

$y = e^{ax}$ $(a = 1)$

Folium of Descartes

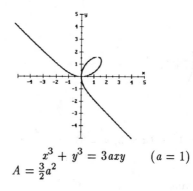

$x^3 + y^3 = 3axy$ $(a = 1)$

$A = \dfrac{3}{2}a^2$

Four-leaved rose

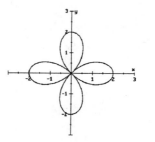

$r = a\cos 2\theta$ $(a = 2)$

Gamma function

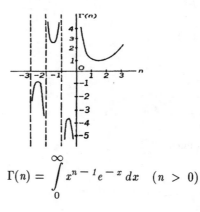

$$\Gamma(n) = \int_0^\infty x^{n-1}e^{-x}\,dx \quad (n > 0)$$

Hyperbola

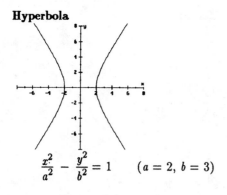

$$\frac{x^2}{a^2} - \frac{y^2}{b^2} = 1 \qquad (a = 2,\ b = 3)$$

Hyperbolic secant

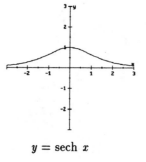

$$y = \operatorname{sech} x$$

Hyperbolic cosecant

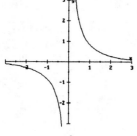

$$y = \operatorname{csch} x$$

Hyperbolic sine

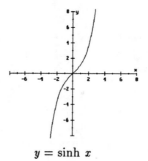

$$y = \sinh x$$

Hyperbolic cosine (Catenary)

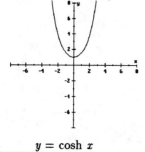

$$y = \cosh x$$

Hyperbolic tangent

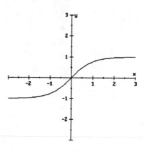

$$y = \tanh x$$

Hyperbolic cotangent

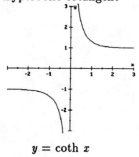

$$y = \coth x$$

Hypocycloid

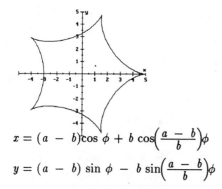

$$x = (a - b)\cos \phi + b \cos\left(\frac{a - b}{b}\right)\phi$$

$$y = (a - b) \sin \phi - b \sin\left(\frac{a - b}{b}\right)\phi$$

This curve is described by a point P on a circle of radius b as it rolls on the inside of a circle of radius a.

Hypocycloid with four cusps (see Astroid)
Hypocycloid with three cusps (see Deltoid)

Identity function

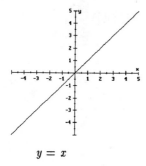

$$y = x$$

Inverse cosecant

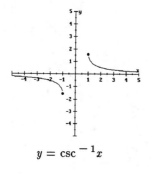

$$y = \csc^{-1}x$$

Inverse cosine

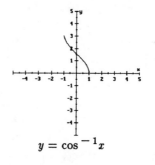

$$y = \cos^{-1}x$$

Inverse cotangent

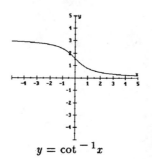

$$y = \cot^{-1}x$$

Inverse secant

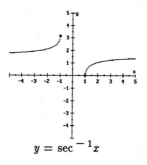

$$y = \sec^{-1}x$$

Inverse sine

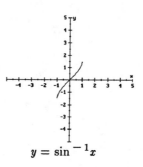

$$y = \sin^{-1}x$$

Inverse tangent

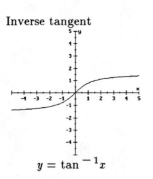

$$y = \tan^{-1}x$$

Involute of a circle

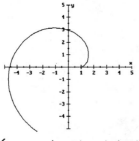

$$\begin{cases} x = a(\cos\phi + \phi\sin\phi) \\ y = a(\sin\phi - \phi\cos\phi) \end{cases} \qquad (a = 1)$$

Lemniscate (or Lemniscate of Bernoulli)

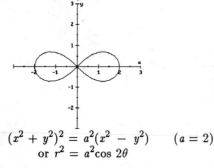

$$(x^2 + y^2)^2 = a^2(x^2 - y^2) \qquad (a = 2)$$
$$\text{or } r^2 = a^2\cos 2\theta$$

Limaçon of Pascal

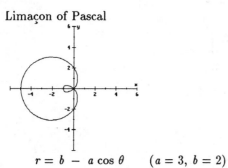

$$r = b - a\cos\theta \qquad (a = 3,\ b = 2)$$

Lituus

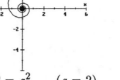

$$r^2\theta = a^2 \qquad (a = 2)$$

Logarithmic curve

$$y = \log_a x \qquad (a = 10)$$

Ouija board curve (see cochleoid)

Ovals of Cassini

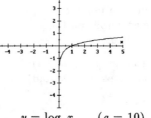

$$b > k \qquad (b = 2,\ k = 1)$$

$$b < k \qquad (b = 2,\ k = 5)$$
$$(x^2 + y^2 + b^2)^2 - 4b^2x^2 = k^2$$

These curves are sections of a torus on planes parallel to the axis of the torus.

Parabola

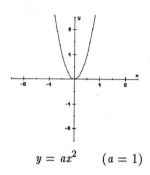

$$y = ax^2 \qquad (a = 1)$$

Probability curve

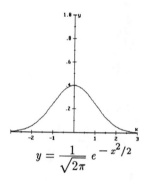

$$y = \frac{1}{\sqrt{2\pi}}\, e^{-x^2/2}$$

Prolate cycloid (see Trochoid)

Quadratrix of Hippias

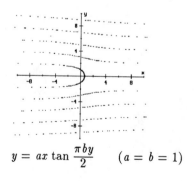

$$y = ax \tan \frac{\pi by}{2} \qquad (a = b = 1)$$

Reciprocal function

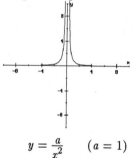

$$y = \frac{a}{x} \qquad (a = 1)$$

Reciprocal squared function

$$y = \frac{a}{x^2} \qquad (a = 1)$$

Rose curves (indexed under number of leaves)

Secant function

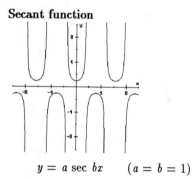

$$y = a \sec bx \qquad (a = b = 1)$$

Semicubical parabola

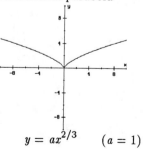

$$y = ax^{2/3} \qquad (a = 1)$$

Serpentine curve

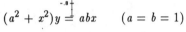

$$(a^2 + x^2)y = abx \qquad (a = b = 1)$$

Sine function

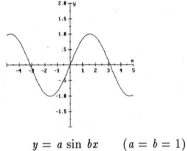

$$y = a \sin bx \qquad (a = b = 1)$$

Sinusoid curve (general sine function)
$$y = a \sin(bx + c)$$

Spiral of Archimedes

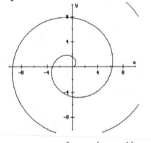

$$r = a\theta \qquad (a = 1)$$

Spiral, hyperbolic

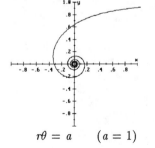

$$r\theta = a \qquad (a = 1)$$

Spiral, logarithmic

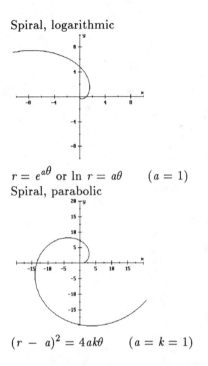

$$r = e^{a\theta} \text{ or } \ln r = a\theta \qquad (a = 1)$$

Spiral, parabolic

$$(r - a)^2 = 4ak\theta \qquad (a = k = 1)$$

Square root function

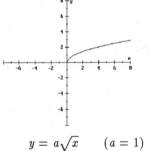

$$y = a\sqrt{x} \qquad (a = 1)$$

Strophoid

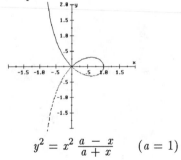

$$y^2 = x^2 \frac{a - x}{a + x} \qquad (a = 1)$$

Tangent function

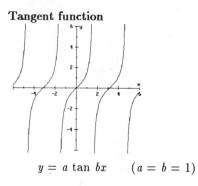

$$y = a \tan bx \qquad (a = b = 1)$$

Three-leaved rose

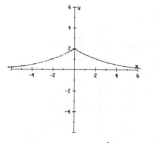

$$r = a \cos 3\theta \qquad (a = 4)$$

Tractrix

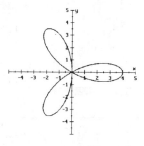

$$x = a(\ln \cot \tfrac{1}{2}\phi - \cos \phi)$$
$$y = a \sin \phi \quad (a = 2)$$

Trochoid

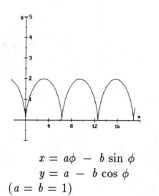

$$x = a\phi - b \sin \phi$$
$$y = a - b \cos \phi$$
$$(a = b = 1)$$

Two-leaved rose (see lemniscate)

Witch of Agnesi

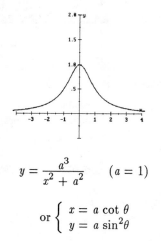

$$y = \frac{a^3}{x^2 + a^2} \qquad (a = 1)$$

$$\text{or} \begin{cases} x = a \cot \theta \\ y = a \sin^2\theta \end{cases}$$

CHAPTER 8

Limit Formulas

8.1

Definition of Limit

LIMIT OF A FUNCTION
(INFORMAL DEFINITION)

The notation
$$\lim_{x \to c} f(x) = L$$

is read "the limit of $f(x)$ as x approaches c is L" and means that the functional values $f(x)$ can be made arbitrarily close to L by choosing x sufficiently close to c.

LIMIT OF A FUNCTION
(FORMAL DEFINITION)

The limit statement
$$\lim_{x \to c} f(x) = L$$
means that for each $\epsilon > 0$, there corresponds a number $\delta > 0$ with the property that
$$|f(x) - L| < \epsilon \quad \text{whenever} \quad 0 < |x - c| < \delta$$

A FUNCTION
DIVERGES TO
INFINITY
(INFORMAL DEFINITION)

A function f that increases or decreases without bound as x approaches c is said to **diverge to infinity** (∞) at c. We indicate this behavior by writing
$$\lim_{x \to c} f(x) = +\infty$$
if x increases without bound, and by

$$\lim_{x \to c} f(x) = -\infty$$

if it decreases without bound.

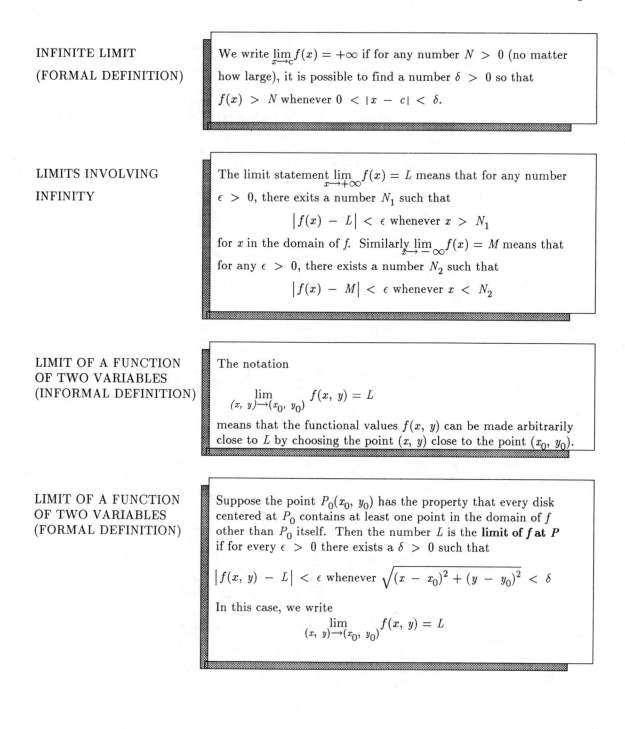

INFINITE LIMIT

(FORMAL DEFINITION)

We write $\lim_{x \to c} f(x) = +\infty$ if for any number $N > 0$ (no matter how large), it is possible to find a number $\delta > 0$ so that $f(x) > N$ whenever $0 < |x - c| < \delta$.

LIMITS INVOLVING

INFINITY

The limit statement $\lim_{x \to +\infty} f(x) = L$ means that for any number $\epsilon > 0$, there exits a number N_1 such that

$$|f(x) - L| < \epsilon \text{ whenever } x > N_1$$

for x in the domain of f. Similarly $\lim_{x \to -\infty} f(x) = M$ means that for any $\epsilon > 0$, there exists a number N_2 such that

$$|f(x) - M| < \epsilon \text{ whenever } x < N_2$$

LIMIT OF A FUNCTION
OF TWO VARIABLES
(INFORMAL DEFINITION)

The notation

$$\lim_{(x,\, y) \to (x_0,\, y_0)} f(x, y) = L$$

means that the functional values $f(x, y)$ can be made arbitrarily close to L by choosing the point (x, y) close to the point (x_0, y_0).

LIMIT OF A FUNCTION
OF TWO VARIABLES
(FORMAL DEFINITION)

Suppose the point $P_0(x_0, y_0)$ has the property that every disk centered at P_0 contains at least one point in the domain of f other than P_0 itself. Then the number L is the **limit of f at P** if for every $\epsilon > 0$ there exists a $\delta > 0$ such that

$$|f(x, y) - L| < \epsilon \text{ whenever } \sqrt{(x - x_0)^2 + (y - y_0)^2} < \delta$$

In this case, we write

$$\lim_{(x,\, y) \to (x_0,\, y_0)} f(x, y) = L$$

8.2
Rules of Limits

BASIC RULES

For any real numbers a and c, suppose the functions f and g both have limits at $x = c$. Also assume that both $\lim_{x \to +\infty} f(x)$ and $\lim_{x \to -\infty} f(x)$ exist.

Limit of a constant $\lim_{x \to c} k = k$ for any constant k

Limit of x $\qquad \lim_{x \to c} x = c$

Scalar rule $\qquad \lim_{x \to c}[af(x)] = a \lim_{x \to c} f(x)$

Sum rule $\qquad \lim_{x \to c}[f(x) + g(x)] = \lim_{x \to c} f(x) + \lim_{x \to c} g(x)$

Difference rule $\qquad \lim_{x \to c}[f(x) - g(x)] = \lim_{x \to c} f(x) - \lim_{x \to c} g(x)$

Linearity rule $\qquad \lim_{x \to +\infty}[af(x) + bg(x)] = a \lim_{x \to +\infty} f(x) + b \lim_{x \to +\infty} g(x)$

Product rules $\qquad \lim_{x \to c}[f(x)g(x)] = [\lim_{x \to c} f(x)][\lim_{x \to c} g(x)]$

$$\lim_{x \to +\infty}[f(x)g(x)] = \left[\lim_{x \to +\infty} f(x)\right]\left[\lim_{x \to +\infty} g(x)\right]$$

Quotient rules $\qquad \lim_{x \to c} \dfrac{f(x)}{g(x)} = \dfrac{\lim_{x \to c} f(x)}{\lim_{x \to c} g(x)}$ if $\lim_{x \to c} g(x) \neq 0$

$$\lim_{x \to +\infty} \frac{f(x)}{g(x)} = \frac{\lim_{x \to +\infty} f(x)}{\lim_{x \to +\infty} g(x)} \text{ if } \lim_{x \to +\infty} g(x) \neq 0$$

Power rules $\qquad \lim_{x \to c}[f(x)]^n = \left[\lim_{x \to c} f(x)\right]^n \quad n$ is a rational number

$$\lim_{x \to +\infty}[f(x)]^n = \left[\lim_{x \to +\infty} f(x)\right]^n$$

Limit limitation theorem

Suppose $\lim_{x \to c} f(x)$ exists and $f(x) \geq 0$ throughout an open interval containing the number c, except possibly at c itself. Then $\lim_{x \to c} f(x) \geq 0$

The squeeze rule

If $g(x) \leq f(x) \leq h(x)$ for all x in an open interval containing c (except possibly at c itself) and if

$$\lim_{x \to c} g(x) = \lim_{x \to c} h(x) = L$$

then $\lim_{x \to c} f(x) = L$

Limits to infinity $\quad \lim_{x \to +\infty} \dfrac{A}{x^n} = 0 \quad$ and $\quad \lim_{x \to -\infty} \dfrac{A}{x^n} = 0$

Infinite limit theorem

If $\lim_{x \to c} f(x) = +\infty$ and $\lim_{x \to c} g(x) = A$, then

$$\lim_{x \to c}[f(x)g(x)] = +\infty \text{ and } \lim_{x \to c} \frac{f(x)}{g(x)} = +\infty \text{ if } A > 0$$

$$\lim_{x \to c}[f(x)g(x)] = -\infty \text{ and } \lim_{x \to c} \frac{f(x)}{g(x)} = -\infty \text{ if } A < 0$$

l'Hôpital's rule

Let f and g be differentiable functions on an open interval containing c (except possibly at c itself).

If $\lim_{x \to c} \dfrac{f(x)}{g(x)}$ produces an indeterminate form $\dfrac{0}{0}$ or $\dfrac{\infty}{\infty}$ then

$$\lim_{x \to c} \frac{f(x)}{g(x)} = \lim_{x \to c} \frac{f'(x)}{g'(x)}$$

provided the limit on the right side exists (or is infinite).

TRIGONOMETRIC

$\lim_{x \to c} \cos x = \cos c$	$\lim_{x \to c} \sec x = \sec c$
$\lim_{x \to c} \sin x = \sin c$	$\lim_{x \to c} \csc x = \csc c$
$\lim_{x \to c} \tan x = \tan c$	$\lim_{x \to c} \cot x = \cot c$

$$\lim_{x \to 0} \frac{\sin x}{x} = 1 \quad \lim_{x \to 0} \frac{\sin ax}{x} = a \quad \lim_{x \to 0} \frac{\tan x}{x} = 1 \quad \lim_{x \to 0} \frac{1 - \cos x}{x} = 0$$

MISCELLANEOUS LIMITS

$$\lim_{n \to +\infty} \left(1 + \frac{1}{n}\right)^n = e \qquad\qquad \lim_{n \to 0} (1 + n)^{1/n} = e$$

$$\lim_{n \to +\infty} \left(1 + \frac{k}{n}\right)^n = e^k \qquad\qquad \lim_{n \to +\infty} P\left(1 + \frac{1}{n}\right)^{nt} = Pe^t$$

$$\lim_{n \to +\infty} n^{1/n} = 1$$

8.3

Limits of a Function of Two Variables

BASIC FORMULAS
AND RULES FOR
LIMITS OF A FUNCTION
OF TWO VARIABLES

Suppose $\lim\limits_{(x,\ y)\to(x_0,\ y_0)} f(x,\ y)$ and $\lim\limits_{(x,\ y)\to(x_0,\ y_0)} g(x,\ y)$ both

exist with $\lim\limits_{(x,\ y)\to(x_0,\ y_0)} f(x,\ y) = L$ and $\lim\limits_{(x,\ y)\to(x_0,\ y_0)} g(x,\ y) = M$

Then,

Scalar rule $\quad \lim\limits_{(x,\ y)\to(x_0,\ y_0)} [af(x,\ y)]$

$$= a \lim\limits_{(x,\ y)\to(x_0,\ y_0)} f(x,\ y) = aL$$

Sum rule $\quad \lim\limits_{(x,\ y)\to(x_0,\ y_0)} [f + g](x,\ y)$

$$= \left[\lim\limits_{(x,\ y)\to(x_0,\ y_0)} f(x,\ y) \right] + \left[\lim\limits_{(x,\ y)\to(x_0,\ y_0)} g(x,\ y) \right]$$

$$= L + M$$

Product rule $\quad \lim\limits_{(x,\ y)\to(x_0,\ y_0)} [fg](x,\ y)$

$$= \left[\lim\limits_{(x,\ y)\to(x_0,\ y_0)} f(x,\ y) \right] \left[\lim\limits_{(x,\ y)\to(x_0,\ y_0)} g(x,\ y) \right]$$

$$= LM$$

Quotient rule $\quad \lim\limits_{(x,\ y)\to(x_0,\ y_0)} \left[\dfrac{f}{g} \right](x,\ y) = \dfrac{\lim\limits_{(x,\ y)\to(x_0,\ y_0)} f(x,\ y)}{\lim\limits_{(x,\ y)\to(x_0,\ y_0)} g(x,\ y)} = \dfrac{L}{M}$

$$\text{if } M \neq 0$$

Substitution rule
If $f(x,\ y)$ is a polynomial or a rational function, limits may be found by substituting for x and y (excluding values that cause division by 0).

CHAPTER 9

Differentiation Formulas

9.1

Definition of Derivative

DERIVATIVE

> The **derivative** of f at x is given by
>
> $$f'(x) = \lim_{\Delta x \to 0} \frac{f(x + \Delta x) - f(x)}{\Delta x}$$
>
> provided this limit exists.

EXAMPLE

Derivative using the definition

Differentiate $f(t) = \sqrt{t}$.

Solution

$$f(t) = \lim_{\Delta t \to 0} \frac{f(t + \Delta t) - f(t)}{\Delta t}$$

$$= \lim_{\Delta t \to 0} \frac{\sqrt{t + \Delta t} - \sqrt{t}}{\Delta t}$$

$$= \lim_{\Delta t \to 0} \frac{\sqrt{t + \Delta t} - \sqrt{t}}{\Delta t}\left(\frac{\sqrt{t + \Delta t} + \sqrt{t}}{\sqrt{t + \Delta t} + \sqrt{t}}\right)$$

$$= \lim_{\Delta t \to 0} \frac{(t + \Delta t) - t}{\Delta t(\sqrt{t + \Delta t} + \sqrt{t})} \qquad \textit{Rationalize numerator}$$

$$= \lim_{\Delta t \to 0} \frac{1}{\sqrt{t + \Delta t} + \sqrt{t}}$$

$$= \frac{1}{2\sqrt{t}} \qquad\qquad\qquad \textit{for } t > 0 \qquad\qquad \square$$

PARTIAL DERIVATIVE
OF A FUNCTION OF
TWO VARIABLES

If $z = f(x, y)$, then the **first partial derivative** of f with respect to x and y are the functions f_x and f_y, respectively, defined by

$$f_x(x, y) = \lim_{\Delta x \to 0} \frac{f(x + \Delta x, y) - f(x, y)}{\Delta x}$$

$$f_y(x, y) = \lim_{\Delta y \to 0} \frac{f(x, y + \Delta y) - f(x, y)}{\Delta y}$$

provided the limits exist.

DIRECTIONAL
DERIVATIVE

Let $f(x, y)$ be a function defined on a disk centered at the point $P_0(x_0, y_0)$, and let $\mathbf{u} = u_1\mathbf{i} + u_2\mathbf{j}$ be a unit vector. Then the **directional derivative** of f at P_0 in the direction of $\mathbf{u}$ is given by

$$D_u f(x_0, y_0) = \lim_{h \to 0} \frac{f(x_0 + hu_1, y_0 + hu_2) - f(x_0, y_0)}{h}$$

provided the limit exists.

9.2

Procedural Rules of Differentiation

If f and g are differentiable functions at x, u is a differentiable function of x, and a, b, and c are any real numbers, then the functions cf, $f + g$, fg, and f/g (for $g(x) \neq 0$) are also differentiable and their derivatives satisfy the following formulas:

Name of Rule	*Derivative*
1. **Constant multiple**	$(cf)' = cf'$
2. **Sum rule**	$(f + g)' = f' + g'$
3. **Difference rule**	$(f - g)' = f' - g'$
4. **Linearity rule**	$(af + bg)' = af' + bg'$
5. **Product rule**	$(fg)' = fg' + f'g$
6. **Quotient rule**	$\left(\dfrac{f}{g}\right)' = \dfrac{gf' - fg'}{g^2}$
7. **Chain rule**	$\dfrac{dy}{dx} = \dfrac{dy}{du}\dfrac{du}{dx}$

9.3

Differentiation Rules

EXTENDED POWER RULE

8. $\dfrac{d}{dx} u^n = n u^{n-1} \dfrac{du}{dx}$

TRIGONOMETRIC FUNCTIONS

9. $\dfrac{d}{dx} \cos u = -\sin u \, \dfrac{du}{dx}$ **10.** $\dfrac{d}{dx} \sin u = \cos u \, \dfrac{du}{dx}$

11. $\dfrac{d}{dx} \tan u = \sec^2 u \, \dfrac{du}{dx}$ **12.** $\dfrac{d}{dx} \cot u = -\csc^2 u \, \dfrac{du}{dx}$

13. $\dfrac{d}{dx} \sec u = \sec u \tan u \, \dfrac{du}{dx}$ **14.** $\dfrac{d}{dx} \csc u = -\csc u \cot u \, \dfrac{du}{dx}$

INVERSE TRIGONOMETRIC FUNCTIONS

15. $\dfrac{d}{dx} \cos^{-1} u = \dfrac{-1}{\sqrt{1 - u^2}} \dfrac{du}{dx}$ **16.** $\dfrac{d}{dx} \sin^{-1} u = \dfrac{1}{\sqrt{1 - u^2}} \dfrac{du}{dx}$

17. $\dfrac{d}{dx} \tan^{-1} u = \dfrac{1}{1 + u^2} \dfrac{du}{dx}$ **18.** $\dfrac{d}{dx} \cot^{-1} u = \dfrac{-1}{1 + u^2} \dfrac{du}{dx}$

19. $\dfrac{d}{dx} \sec^{-1} u = \dfrac{1}{|u|\sqrt{u^2 - 1}} \dfrac{du}{dx}$ **20.** $\dfrac{d}{dx} \csc^{-1} u = \dfrac{-1}{|u|\sqrt{u^2 - 1}} \dfrac{du}{dx}$

EXPONENTIAL AND LOGARITHMIC FUNCTIONS

21. $\dfrac{d}{dx} \ln|u| = \dfrac{1}{u} \dfrac{du}{dx}$ **22.** $\dfrac{d}{dx} \log_b|u| = \dfrac{\log_b e}{u} \dfrac{du}{dx}$

$$= \dfrac{1}{u \ln b} \dfrac{du}{dx}$$

23. $\dfrac{d}{dx} e^u = e^u \dfrac{du}{dx}$ **24.** $\dfrac{d}{dx} b^u = b^u \ln b \, \dfrac{du}{dx}$

HYPERBOLIC FUNCTIONS

25. $\dfrac{d}{dx} \cosh u = \sinh u \, \dfrac{du}{dx}$ **26.** $\dfrac{d}{dx} \sinh u = \cosh u \, \dfrac{du}{dx}$

27. $\dfrac{d}{dx} \tanh u = \operatorname{sech}^2 u \, \dfrac{du}{dx}$ **28.** $\dfrac{d}{dx} \coth u = -\operatorname{csch}^2 u \, \dfrac{du}{dx}$

29. $\dfrac{d}{dx} \operatorname{sech} u = -\operatorname{sech} u \tanh u \, \dfrac{du}{dx}$

30. $\dfrac{d}{dx} \operatorname{csch} u = -\operatorname{csch} u \coth u \, \dfrac{du}{dx}$

INVERSE HYPERBOLIC FUNCTIONS

31. $\dfrac{d}{dx} \sinh^{-1} u = \dfrac{1}{\sqrt{u^2 + 1}} \dfrac{du}{dx}$ **32.** $\dfrac{d}{dx} \cosh^{-1} u = \dfrac{1}{\sqrt{u^2 - 1}} \dfrac{du}{dx}$

33. $\dfrac{d}{dx} \tanh^{-1} u = \dfrac{1}{1 - u^2} \dfrac{du}{dx}$ **34.** $\dfrac{d}{dx} \coth^{-1} u = \dfrac{1}{1 - u^2} \dfrac{du}{dx}$

$|u| < 1$ $|u| > 1$

35. $\dfrac{d}{dx} \operatorname{sech}^{-1} u = \dfrac{-1}{u\sqrt{1 - u^2}} \dfrac{du}{dx}$ **36.** $\dfrac{d}{dx} \operatorname{csch}^{-1} u = \dfrac{-1}{|u|\sqrt{1 + u^2}} \dfrac{du}{dx}$

9.4

Functions of Two Variables

Chain Rule for One Independent Variable

Let $f(x, y)$ be a differentiable function of x and y, and let $x = x(t)$ and $y = y(t)$ be differentiable functions of t. Then $z = f(x, y)$ is a differentiable function of t, and

$$\frac{dz}{dt} = \frac{\partial z}{\partial x}\frac{dx}{dt} + \frac{\partial z}{\partial y}\frac{dy}{dt}$$

The Chain Rule for Two Independent Variables

Suppose $z = f(x, y)$ is differentiable at (x, y) and that the partial derivatives of $x = x(u, v)$ and $y = y(u, v)$ exist at (u, v). Then the composite function $z = f[x(u, v), y(u, v)]$ is differentiable at (u, v) with

$$\frac{\partial z}{\partial u} = \frac{\partial z}{\partial x}\frac{\partial x}{\partial u} + \frac{\partial z}{\partial y}\frac{\partial y}{\partial u}$$

and

$$\frac{\partial z}{\partial v} = \frac{\partial z}{\partial x}\frac{\partial x}{\partial v} + \frac{\partial z}{\partial y}\frac{\partial y}{\partial v}$$

GRADIENT

Let $f(x, y)$ have partial derivatives $f_x(x, y)$ and $f_y(x, y)$. Then the **gradient** of f, denoted by ∇f, is given by

$$\nabla f(x, y) = f_x(x, y)\mathbf{i} + f_y(x, y)\mathbf{j}$$

Basic properties of the gradient

Assuming the existence of the following gradients,

constant rule	$\nabla c = 0$ for any constant c
linearity rule	$\nabla(af + bg) = a\nabla f + b\nabla g$ for constants a and b
product rule	$\nabla(fg) = f(\nabla g) + g(\nabla f)$
quotient rule	$\nabla\left(\dfrac{f}{g}\right) = \dfrac{g(\nabla f) - f(\nabla g)}{g^2}$ $g \neq 0$
power rule	$\nabla(f^n) = nf^{n-1}\nabla f$

CHAPTER 10

Integration Formulas

10.1

Definition of Integral

DEFINITE INTEGRAL

> If f is defined on the closed interval $[a, b]$ we say f is **integrable** on $[a, b]$ if
> $$I = \lim_{\|P\| \to 0} \sum_{k=1}^{n} f(\overset{*}{x}_k)\Delta x_k$$
> exists. This limit, if it exists, is called the **definite integral** of f from a to b. The definite integral is denoted by
>
> $$I = \int_a^b f(x)\, dx$$

EXAMPLE

Evaluating a definite integral using the definition

Evaluate $\displaystyle\int_{-2}^{1} 4x\, dx$

Solution

The integral exists since $f(x) = 4x$ is continuous on $[-2, 1]$. Since the integral can be computed by any partition whose norm approaches 0, we shall simplify matters by choosing a partition in which the points are evenly spaced. Specifically, we divide the interval $[-2, 1]$ into n subintervals, each of width

$$\Delta x = \frac{1 - (-2)}{n} = \frac{3}{n}$$

For each k, we choose the kth subinterval representative to be the right endpoint of the kth subinterval; that is,

$$\overset{*}{x}_k = -2 + k\Delta x = -2 + k(\tfrac{3}{n})$$

Finally, we form the Riemann sum

$$\int_{-2}^{1} 4x \, dx = \lim_{\|P\| \to 0} \sum_{k=1}^{n} f(\overset{*}{x}_k) \Delta x_k$$

$$= \lim_{n \to +\infty} \sum_{k=1}^{n} 4\left(-2 + \frac{3k}{n}\right)\left(\frac{3}{n}\right) \qquad n \to +\infty \text{ as } \|P\| \to 0$$

$$= \lim_{n \to +\infty} \frac{12}{n^2} \sum_{k=1}^{n} (-2n + 3k)$$

$$= \lim_{n \to +\infty} \frac{12}{n^2} \left(\sum_{k=1}^{n} (-2n) + \sum_{k=1}^{n} 3k \right)$$

$$= \lim_{n \to +\infty} \frac{12}{n^2} \left((-2n)n + 3\left[\frac{n(n+1)}{2}\right] \right)$$

$$= \lim_{n \to +\infty} \frac{12}{n^2} \left(\frac{-4n^2 + 3n^2 + 3n}{2} \right)$$

$$= \lim_{n \to +\infty} \frac{-6n^2 + 18n}{n^2}$$

$$= -6 \qquad\qquad \square$$

INDEFINITE INTEGRAL

If $\frac{dy}{dx} = f(x)$, then y is the function whose derivative is $f(x)$ and is called the **antiderivative** of $f(x)$ or the **indefinite integral** of $f(x)$ and is denoted by

$$\int f(x) \, dx$$

DOUBLE INTEGRAL

If f is defined on a closed, bounded region R in the xy-plane, then the **double integral of f over R** is defined by

$$\int_R \int f(x, y) \, dA = \lim_{\|P\| \to 0} \sum_{k=1}^{n} f(\overset{*}{x}_k, \overset{*}{y}_k) \Delta A_k$$

provided this limit exists. If this limit exists, then we say that f is **integrable** over R.

TRIPLE INTEGRAL

If f is a continuous function defined over a bounded solid region S, then the **triple integral of f over S** is defined to be the limit

$$\int \int_S \int f(x, y, z) \, dV = \lim_{\|P\| \to 0} \sum_{k=1}^{n} f(\overset{*}{x}_k, \overset{*}{y}_k, \overset{*}{z}_k) \Delta V_k$$

provided this limit exists.

10.2

Procedural Rules for Integration

In the following, u, v, and w are functions of x; a, b p, q, and n are any constants, restricted if indicated. Since the derivative of a constant is 0, all indefinite integrals differ by an arbitrary constant. In the following formulas, all constants of integration are omitted but implied. It is assumed in all cases that division by 0 is excluded.

	Name of Rule	*Integral*

Properties of integrals

Constant multiple
$$\int cf(u)\ du = c\int f(u)\ du$$

Sum rule
$$\int [f(u)+g(u)]\,du = \int f(u)\,du + \int g(u)\,du$$

Difference rule
$$\int [f(u)-g(u)]\,du = \int f(u)\,du - \int g(u)\,du$$

Linearity rule
$$\int [af(u)+bg(u)]\,du = a\int f(u)\,du + b\int g(u)\,du$$

By parts
$$\int u\ dv = uv - \int v\ du$$

Properties of double integrals

Assume that all the given integrals exist.

Linearity rules   For constants a and b

$$\int_D \int [af(x,\,y) + bg(x,\,y)]\ dA = a\int_D \int f(x,\,y)\ dA + b\int_D \int g(x,\,y)\ dA$$

$$\int_S \int \int [af(x,\,y,\,z) + bg(x,\,y,\,z)]\ dV$$
$$= a\int \int_S \int f(x,\,y,\,z)\ dV + b\int \int_S \int g(x,\,y,\,z)\ dV$$

Dominance rules   If $f(x,\,y) \geq g(x,\,y)$ throughout a region D, then

$$\int_D \int f(x,\,y)\ dA \geq \int_D \int g(x,\,y)\ dA$$

If $f(x,\,y,\,z) \geq g(x,\,y,\,z)$ on S, then

$$\int \int_S \int f(x,\,y,\,z)\ dV \geq \int \int_S \int g(x,\,y,\,z)\ dV$$

Subdivision rules

If the region of integration D can be subdivided into two subregions D_1 and D_2, then

$$\int_D \int f(x,\,y)\ dA = \int_{D_1} \int f(x,\,y)\ dA + \int_{D_2} \int f(x,\,y)\ dA$$

If the solid region of integration S can be subdivided into two solid subregions S_1 and S_2, then

$$\int\int\limits_{S}\int f(x,\ y,\ z)\ dV = \int\int\limits_{S_1}\int f(x,\ y,\ z)\ dV + \int\int\limits_{S_2}\int f(x,\ y,\ z)\ dV$$

10.3

Integration Rules

BASIC FORMULAS

1. Constant rule
$$\int 0\ du = c$$

2. Power rule
$$\int u^n\ du = \frac{u^{n+1}}{n+1};\ n \neq -1$$
$$\int u^n\ du = \ln|u|;\quad n = -1$$

3. Exponential rule
$$\int e^u\ du = e^u$$

4. Logarithmic rule
$$\int \ln|u|\ du = u\ln|u| - u$$

Trigonometric rules

5. $\displaystyle\int \sin u\ du = -\cos u$ **6.** $\displaystyle\int \cos u\ du = \sin u$

7. $\displaystyle\int \tan u\ du = -\ln|\cos u|$ **8.** $\displaystyle\int \cot u\ du = \ln|\sin u|$

9. $\displaystyle\int \sec u\ du = \ln|\sec u + \tan u|$ **10.** $\displaystyle\int \csc u\ du = \ln|\csc u - \cot u|$

11. $\displaystyle\int \sec^2 u\ du = \tan u$ **12.** $\displaystyle\int \csc^2 u\ du = -\cot u$

13. $\displaystyle\int \sec u \tan u\ du = \sec u$ **14.** $\displaystyle\int \csc u \cot u\ du = -\csc u$

Exponential rule (base b)

15. $\displaystyle\int b^u\ du = \frac{b^u}{\ln b}\quad b > 0,\ b \neq 1$

Hyperbolic rules

16. $\displaystyle\int \cosh u\ du = \sinh u$ **17.** $\displaystyle\int \sinh u\ du = \cosh u$

18. $\displaystyle\int \tanh u\ du = \ln \cosh u$ **19.** $\displaystyle\int \coth u\ du = \ln|\sinh u|$

20. $\displaystyle\int \operatorname{sech} u\ du = \tan^{-1}(\sinh u)$ **21.** $\displaystyle\int \operatorname{csch} u\ du = -\ln\left|\tanh \frac{u}{2}\right|$

Inverse rules

22. $\displaystyle\int \frac{du}{\sqrt{a^2 - u^2}} = \sin^{-1}\frac{u}{a}$

23. $\displaystyle\int \frac{du}{\sqrt{u^2 - a^2}} = \cosh^{-1}\frac{u}{a}$

24. $\displaystyle\int \frac{du}{a^2 + u^2} = \frac{1}{a}\tan^{-1}\frac{u}{a}$

25. $\displaystyle\int \frac{du}{a^2 - u^2} = \begin{cases} \frac{1}{a}\tanh^{-1}\frac{u}{a}, & \text{if } \left|\frac{u}{a}\right| < 1 \\ \frac{1}{a}\coth^{-1}\frac{u}{a}, & \text{if } \left|\frac{u}{a}\right| > 1 \end{cases}$

26. $\displaystyle\int \frac{du}{u\sqrt{u^2 - a^2}} = \frac{1}{a}\sec^{-1}\left|\frac{u}{a}\right|$

27. $\displaystyle\int \frac{du}{u\sqrt{a^2 - u^2}} = -\frac{1}{a}\operatorname{sech}^{-1}\left|\frac{u}{a}\right|$

28. $\displaystyle\int \frac{du}{\sqrt{1 + u^2}} = \sinh^{-1}u$

29. $\displaystyle\int \frac{du}{u\sqrt{1 + u^2}} = -\operatorname{csch}^{-1}|u|$

Commonly used Reduction Formulas (from the table of integrals)

Cosine squared integration formula $\displaystyle\int \cos^2 u\, du = \frac{1}{2}u + \frac{1}{4}\sin 2u$ **(317)**

Sine squared integration formula $\displaystyle\int \sin^2 u\, du = \frac{1}{2}u - \frac{1}{4}\sin 2u$ **(348)**

Secant integration formula **(428)**

$$\int \sec^n u\, du = \frac{\sec^{n-2}u \tan u}{n - 1} + \frac{n - 2}{n - 1}\int \sec^{n-2}u\, du, \quad n \neq 1$$

Tangent reduction formula **(406)**

$$\int \tan^n u\, du = \frac{\tan^{n-1}u}{n - 1} - \int \tan^{n-2}u\, du$$

10.4

Table of Integrals

INTEGRALS INVOLVING $au + b$

30. $\displaystyle\int (au + b)^n du = \frac{(au + b)^{n+1}}{(n + 1)a}$

31. $\displaystyle\int u(au + b)^n du = \frac{(au + b)^{n+2}}{(n + 2)a^2} - \frac{b(au + b)^{n+1}}{(n + 1)a^2}$

32. $\displaystyle\int u^2(au + b)^n du = \frac{(au + b)^{n+3}}{(n + 3)a^3} - \frac{2b(au + b)^{n+2}}{(n + 2)a^3} + \frac{b^2(au + b)^{n+1}}{(n + 1)a^3}$

33. $\displaystyle\int u^m(au + b)^n du = \begin{cases} \dfrac{u^{m+1}(au + b)^n}{m + n + 1} + \dfrac{nb}{m + n + 1}\displaystyle\int u^m(au + b)^{n-1} du \\[4mm] \dfrac{u^m(au + b)^{n+1}}{(m + n + 1)a} - \dfrac{mb}{(m + n + 1)a}\displaystyle\int u^{m-1}(au + b)^n du \\[4mm] \dfrac{-u^{m+1}(au + b)^{n+1}}{(n + 1)b} + \dfrac{m + n + 2}{(n + 1)b}\displaystyle\int u^m(au + b)^{n+1} du \end{cases}$

34. $\displaystyle\int \frac{du}{au + b} = \frac{1}{a}\ln|au + b|$

35. $\displaystyle\int \frac{u\,du}{au + b} = \frac{u}{a} - \frac{b}{a^2}\ln|au + b|$

36. $\displaystyle\int \frac{u^2\,du}{au + b} = \frac{(au + b)^2}{2a^3} - \frac{2b(au + b)}{a^3} + \frac{b^2}{a^3}\ln|au + b|$

37. $\displaystyle\int \frac{u^3\,du}{au + b} = \frac{(au + b)^3}{3a^4} - \frac{3b(au + b)^2}{2a^4} + \frac{3b^2(au + b)}{a^4} - \frac{b^3}{a^4}\ln|au + b|$

38. $\displaystyle\int \frac{du}{u(au + b)} = \frac{1}{b}\ln\left|\frac{u}{au + b}\right|$

39. $\displaystyle\int \frac{du}{u^2(au + b)} = -\frac{1}{bu} + \frac{a}{b^2}\ln\left|\frac{au + b}{u}\right|$

40. $\displaystyle\int \frac{du}{u^3(au + b)} = \frac{2au - b}{2b^2 u^2} + \frac{a^2}{b^3}\ln\left|\frac{u}{au + b}\right|$

41. $\displaystyle\int \frac{du}{(au + b)^2} = \frac{-1}{a(au + b)}$

42. $\displaystyle\int \frac{u\,du}{(au + b)^2} = \frac{b}{a^2(au + b)} + \frac{1}{a^2}\ln|au + b|$

43. $\displaystyle\int \frac{u^2\,du}{(au + b)^2} = \frac{au + b}{a^3} - \frac{b^2}{a^3(au + b)} - \frac{2b}{a^3}\ln|au + b|$

44. $\displaystyle\int \frac{u^3\,du}{(au+b)^2} = \frac{(au+b)^2}{2a^4} - \frac{3b(au+b)}{a^4} + \frac{b^3}{a^4(au+b)} + \frac{3b^2}{a^4}\ln|au+b|$

45. $\displaystyle\int \frac{du}{u(au+b)^2} = \frac{1}{b(au+b)} + \frac{1}{b^2}\ln\left|\frac{u}{au+b}\right|$

46. $\displaystyle\int \frac{du}{u^2(au+b)^2} = \frac{-a}{b^2(au+b)} + \frac{1}{b^2u} + \frac{2a}{b^3}\ln\left|\frac{au+b}{u}\right|$

47. $\displaystyle\int \frac{du}{u^3(au+b)^2} = -\frac{(au+b)^2}{2b^4u^2} + \frac{3a(au+b)}{b^4u} - \frac{a^3u}{b^4(au+b)} - \frac{3a^2}{b^4}\ln\left|\frac{au+b}{u}\right|$

48. $\displaystyle\int \frac{du}{(au+b)^3} = \frac{-1}{2a(au+b)^2}$

49. $\displaystyle\int \frac{u\,du}{(au+b)^3} = \frac{-1}{a^2(au+b)} + \frac{b}{2a^2(au+b)^2}$

50. $\displaystyle\int \frac{u^2\,du}{(au+b)^3} = \frac{2b}{a^3(au+b)} - \frac{b^2}{2a^3(au+b)^2} + \frac{1}{a^3}\ln|au+b|$

51. $\displaystyle\int \frac{u^3\,du}{(au+b)^3} = \frac{u}{a^3} - \frac{3b^2}{a^4(au+b)} + \frac{b^3}{2a^4(au+b)^2} - \frac{3b}{a^4}\ln|au+b|$

52. $\displaystyle\int \frac{du}{u(au+b)^3} = \frac{a^2u^2}{2b^3(au+b)^2} - \frac{2au}{b^3(au+b)} - \frac{1}{b^3}\ln\left|\frac{au+b}{u}\right|$

53. $\displaystyle\int \frac{du}{u^2(au+b)^3} = \frac{-a}{2b^2(au+b)^2} - \frac{2a}{b^3(au+b)} - \frac{1}{b^3u} + \frac{3a}{b^4}\ln\left|\frac{au+b}{u}\right|$

54. $\displaystyle\int \frac{du}{u^3(au+b)^3} = \frac{a^4u^2}{2b^5(au+b)^2} - \frac{4a^3u}{b^5(au+b)} - \frac{(au+b)^2}{2b^5u^2} - \frac{6a^2}{b^5}\ln\left|\frac{au+b}{u}\right|$

INTEGRALS INVOLVING $u^2 + a^2$

---

55. $\displaystyle\int \frac{du}{u^2+a^2} = \frac{1}{a}\tan^{-1}\frac{u}{a}$

56. $\displaystyle\int \frac{u\,du}{u^2+a^2} = \frac{1}{2}\ln(u^2+a^2)$

57. $\displaystyle\int \frac{u^2\,du}{u^2+a^2} = u - a\tan^{-1}\frac{u}{a}$

58. $\displaystyle\int \frac{u^3\,du}{u^2+a^2} = \frac{u^2}{2} - \frac{a^2}{2}\ln(u^2+a^2)$

59. $\displaystyle\int \frac{du}{u(u^2+a^2)} = \frac{1}{2a^2}\ln\left(\frac{u^2}{u^2+a^2}\right)$

60. $\displaystyle\int \frac{du}{u^2(u^2+a^2)} = -\frac{1}{a^2u} - \frac{1}{a^3}\tan^{-1}\frac{u}{a}$

61. $\displaystyle\int \frac{du}{u^3(u^2 + a^2)} = -\frac{1}{2a^2u^2} - \frac{1}{2a^4}\ln\left(\frac{u^2}{u^2 + a^2}\right)$

62. $\displaystyle\int \frac{du}{(u^2 + a^2)^2} = \frac{u}{2a^2(u^2 + a^2)} + \frac{1}{2a^3}\tan^{-1}\frac{u}{a}$

63. $\displaystyle\int \frac{u\,du}{(u^2 + a^2)^2} = \frac{-1}{2(u^2 + a^2)}$

64. $\displaystyle\int \frac{u^2\,du}{(u^2 + a^2)^2} = \frac{-u}{2(u^2 + a^2)} + \frac{1}{2a^3}\tan^{-1}\frac{u}{a}$

65. $\displaystyle\int \frac{u^3\,du}{(u^2 + a^2)^2} = \frac{a^2}{2(u^2 + a^2)} + \frac{1}{2}\ln(u^2 + a^2)$

66. $\displaystyle\int \frac{du}{u(u^2 + a^2)^2} = \frac{1}{2a^2(u^2 + a^2)} + \frac{1}{2a^4}\ln\left(\frac{u^2}{u^2 + a^2}\right)$

67. $\displaystyle\int \frac{du}{u^2(u^2 + a^2)^2} = -\frac{1}{a^4u} - \frac{u}{2a^4(u^2 + a^2)} - \frac{3}{2a^5}\tan^{-1}\frac{u}{a}$

68. $\displaystyle\int \frac{du}{u^3(u^2 + a^2)^2} = \frac{1}{2a^4u^2} - \frac{1}{2a^4(u^2 + a^2)} - \frac{1}{a^6}\ln\left(\frac{u^2}{u^2 + a^2}\right)$

69. $\displaystyle\int \frac{du}{(u^2 + a^2)^n} = \frac{u}{2(n-1)a^2(u^2 + a^2)^{n-1}} + \frac{2n-3}{(2n-2)a^2}\int \frac{du}{(u^2 + a^2)^{n-1}}$

70. $\displaystyle\int \frac{u\,du}{(u^2 + a^2)^n} = \frac{-1}{2(n-1)(u^2 + a^2)^{n-1}}$

71. $\displaystyle\int \frac{du}{u(u^2 + a^2)^n} = \frac{1}{2(n-1)a^2(u^2 + a^2)^{n-1}} + \frac{1}{a^2}\int \frac{du}{u(u^2 + a^2)^{n-1}}$

72. $\displaystyle\int \frac{u^m du}{(u^2 + a^2)^n} = \int \frac{u^{m-2}\,du}{(u^2 + a^2)^{n-1}} - a^2\int \frac{u^{m-2}\,du}{(u^2 + a^2)^n}$

73. $\displaystyle\int \frac{du}{u^m(u^2 + a^2)^n} = \frac{1}{a^2}\int \frac{du}{u^m(u^2 + a^2)^{n-1}} - \frac{1}{a^2}\int \frac{du}{u^{m-2}(u^2 + a^2)^n}$

INTEGRALS INVOLVING $u^2 - a^2$, $u^2 > a^2$

74. $\displaystyle\int \frac{du}{u^2 - a^2} = \frac{1}{2a}\ln\left|\frac{u-a}{u+a}\right|$ or $-\frac{1}{a}\coth^{-1}\frac{u}{a}$

75. $\displaystyle\int \frac{u\,du}{u^2 - a^2} = \frac{1}{2}\ln\left|u^2 - a^2\right|$

76. $\displaystyle\int \frac{u^2\,du}{u^2-a^2} = u + \frac{a}{2}\ln\left|\frac{u-a}{u+a}\right|$

77. $\displaystyle\int \frac{u^3\,du}{u^2-a^2} = \frac{u^2}{2} + \frac{a^2}{2}\ln|u^2-a^2|$

78. $\displaystyle\int \frac{du}{u(u^2-a^2)} = \frac{1}{2a^2}\ln\left|\frac{u^2-a^2}{u^2}\right|$

79. $\displaystyle\int \frac{du}{u^2(u^2-a^2)} = \frac{1}{a^2 u} + \frac{1}{2a^3}\ln\left|\frac{u-a}{u+a}\right|$

80. $\displaystyle\int \frac{du}{u^3(u^2-a^2)} = \frac{1}{2a^2 u^2} - \frac{1}{2a^4}\ln\left|\frac{u^2}{u^2-a^2}\right|$

81. $\displaystyle\int \frac{du}{(u^2-a^2)^2} = \frac{-u}{2a^2(u^2-a^2)} - \frac{1}{4a^3}\ln\left|\frac{u-a}{u+a}\right|$

82. $\displaystyle\int \frac{u\,du}{(u^2-a^2)^2} = \frac{-1}{2(u^2-a^2)}$

83. $\displaystyle\int \frac{u^2\,du}{(u^2-a^2)^2} = \frac{-u}{2(u^2-a^2)} + \frac{1}{4a}\ln\left|\frac{u-a}{u+a}\right|$

84. $\displaystyle\int \frac{u^3\,du}{(u^2-a^2)^2} = \frac{-a^2}{2(u^2-a^2)} + \frac{1}{2}\ln|u^2-a^2|$

85. $\displaystyle\int \frac{du}{u(u^2-a^2)^2} = \frac{-1}{2a^2(u^2-a^2)} + \frac{1}{2a}\ln\left|\frac{u^2}{u^2-a^2}\right|$

86. $\displaystyle\int \frac{du}{u^2(u^2-a^2)^2} = -\frac{1}{a^4 u} - \frac{u}{2a^4(u^2-a^2)} - \frac{3}{4a^5}\ln\left|\frac{u-a}{u+a}\right|$

87. $\displaystyle\int \frac{du}{u^3(u^2-a^2)^2} = -\frac{1}{2a^4 u^2} - \frac{1}{2a^4(u^2-a^2)} + \frac{1}{a^6}\ln\left|\frac{u^2}{u^2-a^2}\right|$

88. $\displaystyle\int \frac{du}{(u^2-a^2)^n} = \frac{-u}{2(n-1)a^2(u^2-a^2)^{n-1}} - \frac{2n-3}{(2n-2)a^2}\int \frac{du}{(u^2-a^2)^{n-1}}$

89. $\displaystyle\int \frac{u\,du}{(u^2-a^2)^n} = \frac{-1}{2(n-1)(u^2-a^2)^{n-1}}$

90. $\displaystyle\int \frac{du}{u(u^2-a^2)^n} = \frac{-1}{2(n-1)a^2(u^2-a^2)^{n-1}} - \frac{1}{a^2}\int \frac{du}{u(u^2-a^2)^{n-1}}$

91. $\displaystyle\int \frac{u^m\,du}{(u^2-a^2)^n} = \int \frac{u^{m-2}\,du}{(u^2-a^2)^{n-1}} + a^2\int \frac{u^{m-2}\,du}{(u^2-a^2)^n}$

92. $\displaystyle\int \frac{du}{u^m(u^2 - a^2)^n} = \frac{1}{a^2}\int \frac{du}{u^{m-2}(u^2 - a^2)^n} - \frac{1}{a^2}\int \frac{du}{u^m(u^2 - a^2)^{n-1}}$

INTEGRALS INVOLVING $a^2 - u^2$, $u^2 < a^2$

---

93. $\displaystyle\int \frac{du}{a^2 - u^2} = \frac{1}{2a}\ln\left|\frac{a + u}{a - u}\right|$ or $\frac{1}{a}\tanh^{-1}\frac{u}{a}$

94. $\displaystyle\int \frac{u\,du}{a^2 - u^2} = -\frac{1}{2}\ln\left|a^2 - u^2\right|$

95. $\displaystyle\int \frac{u^2\,du}{a^2 - u^2} = -u + \frac{a}{2}\ln\left|\frac{a + u}{a - u}\right|$

96. $\displaystyle\int \frac{u^3\,du}{a^2 - u^2} = -\frac{u^2}{2} - \frac{a^2}{2}\ln\left|a^2 - u^2\right|$

97. $\displaystyle\int \frac{du}{u(a^2 - u^2)} = \frac{1}{2a^2}\ln\left|\frac{u^2}{a^2 - u^2}\right|$

98. $\displaystyle\int \frac{du}{u^2(a^2 - u^2)} = -\frac{1}{a^2 u} + \frac{1}{2a^3}\ln\left|\frac{a + u}{a - u}\right|$

99. $\displaystyle\int \frac{du}{u^3(a^2 - u^2)} = -\frac{1}{2a^2 u^2} + \frac{1}{2a^4}\ln\left|\frac{u^2}{a^2 - u^2}\right|$

100. $\displaystyle\int \frac{du}{(a^2 - u^2)^2} = \frac{u}{2a^2(a^2 - u^2)} + \frac{1}{4a^3}\ln\left|\frac{a + u}{a - u}\right|$

101. $\displaystyle\int \frac{u\,du}{(a^2 - u^2)^2} = \frac{1}{2(a^2 - u^2)}$

102. $\displaystyle\int \frac{u^2\,du}{(a^2 - u^2)^2} = \frac{u}{2(a^2 - u^2)} - \frac{1}{4a}\ln\left|\frac{a + u}{a - u}\right|$

103. $\displaystyle\int \frac{u^3\,du}{(a^2 - u^2)^2} = \frac{a^2}{2(a^2 - u^2)} + \frac{1}{2}\ln\left|a^2 - u^2\right|$

104. $\displaystyle\int \frac{du}{u(a^2 - u^2)^2} = \frac{1}{2a^2(a^2 - u^2)} + \frac{1}{2a^4}\ln\left|\frac{u^2}{a^2 - u^2}\right|$

105. $\displaystyle\int \frac{du}{u^2(a^2 - u^2)^2} = \frac{-1}{a^4 u} + \frac{u}{2a^4(a^2 - u^2)} + \frac{3}{4a^5}\ln\left|\frac{a + u}{a - u}\right|$

106. $\displaystyle\int \frac{du}{u^3(a^2 - u^2)^2} = \frac{-1}{2a^4 u^2} + \frac{1}{2a^4(a^2 - u^2)} + \frac{1}{a^6}\ln\left|\frac{u^2}{a^2 - u^2}\right|$

107. $\displaystyle \int \frac{du}{(a^2 - u^2)^n} = \frac{u}{2(n-1)a^2(a^2 - u^2)^{n-1}} + \frac{2n-3}{(2n-2)a^2} \int \frac{du}{(a^2 - u^2)^{n-1}}$

108. $\displaystyle \int \frac{u\, du}{(a^2 - u^2)^n} = \frac{1}{2(n-1)(a^2 - u^2)^{n-1}}$

109. $\displaystyle \int \frac{du}{u(a^2 - u^2)^n} = \frac{1}{2(n-1)a^2(a^2 - u^2)^{n-1}} + \frac{1}{a^2} \int \frac{du}{u(a^2 - u^2)^{n-1}}$

110. $\displaystyle \int \frac{u^m\, du}{(a^2 - u^2)^n} = a^2 \int \frac{u^{m-2}\,du}{(a^2 - u^2)^n} - \int \frac{u^{m-2}\,du}{(a^2 - u^2)^{n-1}}$

111. $\displaystyle \int \frac{du}{u^m(a^2 - u^2)^n} = \frac{1}{a^2} \int \frac{du}{u^m(a^2 - u^2)^{n-1}} + \frac{1}{a^2} \int \frac{du}{u^{m-2}(a^2 - u^2)^n}$

INTEGRALS INVOLVING $au + b$ AND $pu + q$

112. $\displaystyle \int \frac{du}{(au+b)(pu+q)} = \frac{1}{bp-aq} \ln\left|\frac{pu+q}{au+b}\right|$

113. $\displaystyle \int \frac{u\, du}{(au+b)(pu+q)} = \frac{1}{bp-aq}\left\{ \frac{b}{a} \ln|au+b| - \frac{q}{p} \ln(|u|+q) \right\}$

114. $\displaystyle \int \frac{du}{(au+b)^2(pu+q)} = \frac{1}{bp-aq}\left\{ \frac{1}{au+b} + \frac{p}{bp-aq} \ln\left|\frac{pu+q}{au+b}\right| \right\}$

115. $\displaystyle \int \frac{u\, du}{(au+b)^2(pu+q)} = \frac{1}{bp-aq}\left\{ \frac{q}{bp-aq} \ln\left|\frac{au+b}{pu+q}\right| - \frac{b}{a(au+b)} \right\}$

116. $\displaystyle \int \frac{u^2\, du}{(au+b)^2(pu+q)} = \frac{b^2}{(bp-aq)a^2(au+b)} + \frac{1}{(bp-aq)^2}\left\{ \frac{q^2}{p} \ln|pu+q| + \frac{b(bp-2aq)}{a^2} \ln|au+b| \right\}$

117. $\displaystyle \int \frac{du}{(au+b)^m(pu+q)^n}$

$\displaystyle = \frac{1}{(n-1)(bp-aq)}\left\{ \frac{1}{(au+b)^{m-1}(pu+q)^{n-1}} + a(m+n-2) \int \frac{du}{(au+b)^m(pu+q)^{n-1}} \right\}$

118. $\displaystyle \int \frac{au+b}{pu+q}\, du = \frac{au}{p} + \frac{bp-aq}{p^2} \ln|pu+q|$

119. $\displaystyle \int \frac{(au+b)^m}{(pu+q)^n}\, du$

$$
= \begin{cases}
\dfrac{-1}{(n-1)(bp-aq)} \left\{ \dfrac{(au+b)^{m+1}}{(pu+q)^{n-1}} + (n-m-2)a \displaystyle\int \dfrac{(au+b)^m}{(pu+q)^{n-1}} \, du \right\} \\[4mm]
\dfrac{-1}{(n-m-1)p} \left\{ \dfrac{(au+b)^m}{(pu+q)^{n-1}} + m(bp-aq) \displaystyle\int \dfrac{(au+b)^{m-1}}{(pu+q)^n} \, du \right\} \\[4mm]
\dfrac{-1}{(n-1)p} \left\{ \dfrac{(au+b)^m}{(pu+q)^{n-1}} - ma \displaystyle\int \dfrac{(au+b)^{m-1}}{(pu+q)^{n-1}} \, du \right\}
\end{cases}
$$

INTEGRALS INVOLVING $au^2 + bu + c$

120. $\displaystyle \int \frac{du}{au^2 + bu + c} = \begin{cases} \dfrac{2}{\sqrt{4ac - b^2}} \tan^{-1} \dfrac{2au + b}{\sqrt{4ac - b^2}} \\[4mm] \dfrac{1}{\sqrt{b^2 - 4ac}} \ln \left| \dfrac{2au + b - \sqrt{b^2 - 4ac}}{2au + b + \sqrt{b^2 - 4ac}} \right| \end{cases}$

121. $\displaystyle \int \frac{u \, du}{au^2 + bu + c} = \frac{1}{2a} \ln |au^2 + bu + c| - \frac{b}{2a} \int \frac{du}{au^2 + bu + c}$

122. $\displaystyle \int \frac{u^2 \, du}{au^2 + bu + c} = \frac{u}{a} - \frac{b}{2a^2} \ln |au^2 + bu + c| + \frac{b^2 - 2ac}{2a^2} \int \frac{du}{au^2 + bu + c}$

123. $\displaystyle \int \frac{u^m \, du}{au^2 + bu + c} = \frac{u^{m-1}}{(m-1)a} - \frac{c}{a} \int \frac{u^{m-2} \, du}{au^2 + bu + c} - \frac{b}{a} \int \frac{u^{m-1} \, du}{au^2 + bu + c}$

124. $\displaystyle \int \frac{du}{u(au^2 + bu + c)} = \frac{1}{2c} \ln \left| \frac{u^2}{au^2 + bu + c} \right| - \frac{b}{2c} \int \frac{du}{au^2 + bu + c}$

125. $\displaystyle \int \frac{du}{u^2(au^2 + bu + c)} = \frac{b}{2c^2} \ln \left| \frac{au^2 + bu + c}{u^2} \right| - \frac{1}{cu} + \frac{b^2 - 2ac}{2c^2} \int \frac{du}{au^2 + bu + c}$

126. $\displaystyle \int \frac{du}{u^n(au^2 + bu + c)}$

$$
= -\frac{1}{(n-1)cu^{n-1}} - \frac{b}{c} \int \frac{du}{u^{n-1}(au^2 + bu + c)} - \frac{a}{c} \int \frac{du}{u^{n-2}(au^2 + bu + c)}
$$

127. $\displaystyle\int \frac{du}{(au^2 + bu + c)} = \frac{2au + b}{(4ac - b^2)(au^2 + bu + c)} + \frac{2a}{4ac - b^2} \int \frac{du}{au^2 + bu + c}$

128. $\displaystyle\int \frac{u\, du}{(au^2 + bu + c)^2} = -\frac{bu + 2c}{(4ac - b^2)(au^2 + bu + c)} - \frac{b}{4ac - b^2} \int \frac{du}{au^2 + bu + c}$

129. $\displaystyle\int \frac{u^2\, du}{(au^2 + bu + c)^2} = \frac{(b^2 - 2ac)u + bc}{a(4ac - b^2)(au^2 + bu + c)} + \frac{2c}{4ac - b^2} \int \frac{du}{au^2 + bu + c}$

130. $\displaystyle\int \frac{u^m du}{(au^2 + bu + c)}$

$\displaystyle = \frac{u^{m-1}}{(2n - m - 1)a(au^2 + bu + c)^{n-1}} + \frac{(m - 1)c}{(2n - m - 1)a} \int \frac{u^{m-2}du}{(au^2 + bu + c)^n} - \frac{(n - m)b}{(2n - m - 1)a} \int \frac{u^{m-1}du}{(au^2 + bu + c)^n}$

131. $\displaystyle\int \frac{u^{2n-1}\, du}{(au^2 + bu + c)^n}$

$\displaystyle = -\frac{1}{a} \int \frac{u^{2n-3}\, du}{(au^2 + bu + c)^{n-1}} - \frac{c}{a} \int \frac{u^{2n-3}\, du}{(au^2 + bu + c)^n} - \frac{b}{a} \int \frac{u^{2n-2}\, du}{(au^2 + bu + c)^n}$

132. $\displaystyle\int \frac{du}{u(au^2 + bu + c)^2} = \frac{1}{2c(au^2 + bu + c)} - \frac{b}{2c} \int \frac{du}{(au^2 + bu + c)^2} + \frac{1}{c} \int \frac{du}{u(au^2 + bu + c)}$

133. $\displaystyle\int \frac{du}{u^2(au^2 + bu + c)^2} = -\frac{1}{cu(au^2 + bu + c)} - \frac{3a}{c} \int \frac{du}{(au^2 + bu + c)^2} - \frac{2b}{c} \int \frac{du}{u(au^2 + bu + c)^2}$

134. $\displaystyle\int \frac{du}{u^m(au^2 + bu + c)^n} = -\frac{1}{(m - 1)cu^{m-1}(au^2 + bu + c)^{n-1}}$

$\displaystyle - \frac{(m + 2n - 3)a}{(m - 1)c} \int \frac{du}{u^{m-2}(au^2 + bu + c)^n} - \frac{(m + n - 2)b}{(m - 1)c} \int \frac{du}{u^{m-1}(au^2 + bu + c)^n}$

INTEGRALS INVOLVING $\sqrt{au + b}$

135. $\displaystyle\int \frac{du}{\sqrt{au + b}} = \frac{2\sqrt{au + b}}{a}$

136. $\displaystyle\int \frac{u\,du}{\sqrt{au+b}} = \frac{2(au-2b)}{3a^2}\sqrt{au+b}$

137. $\displaystyle\int \frac{u^2\,du}{\sqrt{au+b}} = \frac{2(3a^2u^2 - 4abu + 8b^2)}{15a^3}\sqrt{au+b}$

138. $\displaystyle\int \frac{du}{u\sqrt{au+b}} = \begin{cases} \dfrac{1}{\sqrt{b}}\ln\left|\dfrac{\sqrt{au+b}-\sqrt{b}}{\sqrt{au+b}+\sqrt{b}}\right| \\[3ex] \dfrac{2}{\sqrt{-b}}\tan^{-1}\sqrt{\dfrac{au+b}{-b}} \end{cases}$

139. $\displaystyle\int \frac{du}{u^2\sqrt{au+b}} = -\frac{\sqrt{au+b}}{bu} - \frac{a}{2b}\int\frac{du}{u\sqrt{au+b}}$

140. $\displaystyle\int \sqrt{au+b}\,du = \frac{2\sqrt{(au+b)^3}}{3a}$

141. $\displaystyle\int u\sqrt{au+b}\,du = \frac{2(3au-2b)}{15a^2}\sqrt{(au+b)^3}$

142. $\displaystyle\int u^2\sqrt{au+b}\,du = \frac{2(15a^2u^2 - 12abu + 8b^2)}{105a^3}\sqrt{(au+b)^3}$

143. $\displaystyle\int \frac{\sqrt{au+b}}{u}\,du = 2\sqrt{au+b} + b\int\frac{du}{u\sqrt{au+b}}$

144. $\displaystyle\int \frac{\sqrt{au+b}}{u^2}\,du = \frac{\sqrt{au+b}}{u} + \frac{a}{2}\int\frac{du}{u\sqrt{au+b}}$

145. $\displaystyle\int \frac{u^m}{\sqrt{au+b}}\,du = \frac{2u^m\sqrt{au+b}}{(2m+1)a} - \frac{2mb}{(2m+1)a}\int\frac{u^{m-1}}{\sqrt{au+b}}\,du$

146. $\displaystyle\int \frac{du}{u^m\sqrt{au+b}} = -\frac{\sqrt{au+b}}{(m-1)bu^{m-1}} - \frac{(2m-3)a}{(2m-2)b}\int\frac{du}{u^{m-1}\sqrt{au+b}}$

147. $\displaystyle\int u^m\sqrt{au+b}\,du = \frac{2u^m}{(2m+3)a}(au+b)^{3/2} - \frac{2mb}{(2m+3)a}\int u^{m-1}\sqrt{au+b}\,du$

148. $\displaystyle\int \frac{\sqrt{au+b}}{u^m}\,du = -\frac{\sqrt{au+b}}{(m-1)u^{m-1}} + \frac{a}{2(m-1)}\int\frac{du}{u^{m-1}\sqrt{au+b}}$

149. $\displaystyle\int \frac{\sqrt{au+b}}{u^m}\,du = \frac{-(au+b)^{3/2}}{(m-1)bu^{m-1}} - \frac{(2m-5)a}{(2m-2)b}\int\frac{\sqrt{au+b}}{u^{m-1}}\,du$

150. $\displaystyle\int (au+b)^{m/2}\,du = \frac{2(au+b)^{(m+2)/2}}{a(m+2)}$

151. $\int u(au+b)^{m/2}\,du = \dfrac{2(au+b)^{(m+4)/2}}{a^2(m+4)} - \dfrac{2b(au+b)^{(m+2)/2}}{a^2(m+2)}$

152. $\int u^2(au+b)^{m/2}\,du = \dfrac{2(au+b)^{(m+6)/2}}{a^3(m+6)} - \dfrac{4b(au+b)^{(m+4)/2}}{a^3(m+4)} + \dfrac{2b^2(au+b)^{(m+2)/2}}{a^3(m+2)}$

153. $\int \dfrac{(au+b)^{m/2}}{u}\,du = \dfrac{2(au+b)^{m/2}}{m} + b\int \dfrac{(au+b)^{(m-2)/2}}{u}\,du$

154. $\int \dfrac{(au+b)^{m/2}}{u^2}\,du = \dfrac{(au+b)^{(m+2)/2}}{bu} + \dfrac{ma}{2b}\int \dfrac{(au+b)^{m/2}}{u}\,du$

155. $\int \dfrac{du}{u(au+b)^{m/2}} = \dfrac{2}{(m-2)b(au+b)^{(m-2)/2}} + \dfrac{1}{b}\int \dfrac{du}{u(au+b)^{(m-2)/2}}$

INTEGRALS INVOLVING $\sqrt{au+b}$ AND $pu+q$

156. $\int \dfrac{pu+q}{\sqrt{au+b}}\,du = \dfrac{2(apu+3aq-2bp)}{3a^2}\sqrt{au+b}$

157. $\int \dfrac{du}{(pu+q)\sqrt{au+b}} = \begin{cases} \dfrac{1}{\sqrt{bp-aq}\sqrt{p}}\ln\left|\dfrac{\sqrt{p(au+b)}-\sqrt{bp-aq}}{\sqrt{p(au+b)}+\sqrt{bp-aq}}\right| \\[4mm] \dfrac{2}{\sqrt{aq-bp}\sqrt{p}}\tan^{-1}\sqrt{\dfrac{p(au+b)}{aq-bp}} \end{cases}$

158. $\int \dfrac{\sqrt{au+b}}{pu+q}\,du = \begin{cases} \dfrac{2\sqrt{au+b}}{p} + \dfrac{\sqrt{bp-aq}}{p\sqrt{p}}\ln\left|\dfrac{\sqrt{p(au+b)}-\sqrt{bp-aq}}{\sqrt{p(au+b)}+\sqrt{bp-aq}}\right| \\[4mm] \dfrac{2\sqrt{au+b}}{p} - \dfrac{2\sqrt{aq-bp}}{p\sqrt{p}}\tan^{-1}\sqrt{\dfrac{p(au+b)}{aq-bp}} \end{cases}$

159. $\int (pu+q)^n\sqrt{au+b}\,du = \dfrac{2(pu+q)^{n+1}\sqrt{au+b}}{(2n+3)p} + \dfrac{bp-aq}{(2n+3)p}\int \dfrac{(pu+q)^n}{\sqrt{au+b}}\,du$

160. $\int \dfrac{du}{(pu+q)^n\sqrt{au+b}} = \dfrac{\sqrt{au+b}}{(n-1)(aq-bp)(pu+q)^{n-1}} + \dfrac{(2n-3)a}{2(n-1)(aq-bp)}\int \dfrac{du}{(pu+q)^{n-1}\sqrt{au+b}}$

161. $\int \dfrac{(pu+q)^n}{\sqrt{au+b}}\,du = \dfrac{2(pu+q)^n\sqrt{au+b}}{(2n+1)a} + \dfrac{2n(aq-bp)}{(2n+1)a}\int \dfrac{(pu+q)^{n-1}\,du}{\sqrt{au+b}}$

162. $\int \dfrac{\sqrt{au+b}}{(pu+q)^n}\,du = \dfrac{-\sqrt{au+b}}{(n-1)p(pu+q)^{n-1}} + \dfrac{a}{2(n-1)p}\int \dfrac{du}{(pu+q)^{n-1}\sqrt{au+b}}$

INTEGRALS INVOLVING $\sqrt{au+b}$ AND $\sqrt{pu+q}$

163. $\displaystyle\int \frac{du}{\sqrt{(au+b)(pu+q)}} = \begin{cases} \dfrac{2}{\sqrt{ap}} \ln\left| \sqrt{a(pu+q)} + \sqrt{p(au+b)} \right| \\[4mm] \dfrac{2}{\sqrt{-ap}} \tan^{-1}\sqrt{\dfrac{-p(au+b)}{a(pu+q)}} \end{cases}$

164. $\displaystyle\int \frac{du}{\sqrt{(au+b)(pu+q)}} = \frac{\sqrt{(au+b)(pu+q)}}{ap} - \frac{bp+aq}{2ap} \int \frac{du}{\sqrt{(au+b)(pu+q)}}$

165. $\displaystyle\int \sqrt{(au+b)(pu+q)}\; du = \frac{2apu+bp+aq}{4ap} \sqrt{(au+b)(pu+q)} - \frac{(bp-aq)^2}{8ap} \int \frac{du}{\sqrt{(au+b)(pu+q)}}$

166. $\displaystyle\int \sqrt{\frac{pu+q}{au+b}}\; du = \frac{\sqrt{(au+b)(pu+q)}}{a} + \frac{aq-bp}{2a} \int \frac{du}{\sqrt{(au+b)(pu+q)}}$

167. $\displaystyle\int \frac{du}{(pu+q)\sqrt{(au+b)(pu+q)}} = \frac{2\sqrt{au+b}}{(aq-bp)\sqrt{pu+q}}$

INTEGRALS INVOLVING $\sqrt{u^2+a^2}$

168. $\displaystyle\int \sqrt{u^2+a^2}\; du = \frac{u\sqrt{u^2+a^2}}{2} + \frac{a^2}{2} \ln\left| u+\sqrt{u^2+a^2} \right|$

169. $\displaystyle\int u\sqrt{u^2+a^2}\; du = \frac{(u^2+a^2)^{3/2}}{3}$

170. $\displaystyle\int u^2\sqrt{u^2+a^2}\; du = \frac{u(u^2+a^2)^{3/2}}{4} - \frac{a^2 u\sqrt{u^2+a^2}}{8} - \frac{a^4}{8} \ln\left| u+\sqrt{u^2+a^2} \right|$

171. $\displaystyle\int u^3\sqrt{u^2+a^2}\; du = \frac{(u^2+a^2)^{5/2}}{5} - \frac{a^2(u^2+a^2)^{3/2}}{3}$

172. $\displaystyle\int \frac{du}{\sqrt{u^2+a^2}} = \ln\left| u+\sqrt{u^2+a^2} \right| \text{ or } \sinh^{-1}\frac{u}{a}$

173. $\displaystyle\int \frac{u\; du}{\sqrt{u^2+a^2}} = \sqrt{u^2+a^2}$

174. $\displaystyle\int \frac{u^2\ du}{\sqrt{u^2 + a^2}} = \frac{u\sqrt{u^2 + a^2}}{2} - \frac{a^2}{2} \ln\left| u + \sqrt{u^2 + a^2} \right|$

175. $\displaystyle\int \frac{u^3\ du}{\sqrt{u^2 + a^2}} = \frac{(u^2 + a^2)^{3/2}}{3} - a^2\sqrt{u^2 + a^2}$

176. $\displaystyle\int \frac{du}{u\sqrt{u^2 + a^2}} = -\frac{1}{a} \ln\left| \frac{a + \sqrt{u^2 + a^2}}{u} \right|$

177. $\displaystyle\int \frac{du}{u^2\sqrt{u^2 + a^2}} = -\frac{\sqrt{u^2 + a^2}}{a^2 u}$

178. $\displaystyle\int \frac{du}{u^3\sqrt{u^2 + a^2}} = -\frac{\sqrt{u^2 + a^2}}{2a^2 u^2} + \frac{1}{2a^3} \ln\left| \frac{a + \sqrt{u^2 + a^2}}{u} \right|$

179. $\displaystyle\int \frac{\sqrt{u^2 + a^2}}{u}\ du = \sqrt{u^2 + a^2} - a \ln\left| \frac{a + \sqrt{u^2 + a^2}}{u} \right|$

180. $\displaystyle\int \frac{\sqrt{u^2 + a^2}}{u^2}\ du = -\frac{\sqrt{u^2 + a^2}}{u} + \ln\left| u + \sqrt{u^2 + a^2} \right|$

181. $\displaystyle\int \frac{\sqrt{u^2 + a^2}}{u^3}\ du = -\frac{\sqrt{u^2 + a^2}}{2u^2} - \frac{1}{2a} \ln\left| \frac{a + \sqrt{u^2 + a^2}}{u} \right|$

182. $\displaystyle\int \frac{du}{(u^2 + a^2)^{3/2}} = \frac{u}{a^2\sqrt{u^2 + a^2}}$

183. $\displaystyle\int \frac{u\ du}{(u^2 + a^2)^{3/2}} = \frac{-1}{\sqrt{u^2 + a^2}}$

184. $\displaystyle\int \frac{u^2\ du}{(u^2 + a^2)^{3/2}} = \frac{-u}{\sqrt{u^2 + a^2}} + \ln\left| u + \sqrt{u^2 + a^2} \right|$

185. $\displaystyle\int \frac{u^3\ du}{(u^2 + a^2)^{3/2}} = \sqrt{u^2 + a^2} + \frac{a^2}{\sqrt{u^2 + a^2}}$

186. $\displaystyle\int \frac{du}{u(u^2 + a^2)^{3/2}} = \frac{1}{a^2\sqrt{u^2 + a^2}} - \frac{1}{a^3} \ln\left| \frac{a + \sqrt{u^2 + a^2}}{u} \right|$

187. $\displaystyle\int \frac{du}{u^2(u^2 + a^2)^{3/2}} = -\frac{\sqrt{u^2 + a^2}}{a^4 u} - \frac{u}{a^4\sqrt{u^2 + a^2}}$

188. $\displaystyle\int \frac{du}{u^3(u^2 + a^2)^{3/2}} = \frac{-1}{2a^2 u^2\sqrt{u^2 + a^2}} - \frac{3}{2a^4\sqrt{u^2 + a^2}} + \frac{3}{2a^5} \ln\left| \frac{a + \sqrt{u^2 + a^2}}{u} \right|$

189. $\int (u^2 + a^2)^{3/2}\, du = \dfrac{u(u^2 + a^2)^{3/2}}{4} + \dfrac{3a^2 u\sqrt{u^2 + a^2}}{8} + \dfrac{3}{8}a^4 \ln\left| u + \sqrt{u^2 + a^2} \right|$

190. $\int u(u^2 + a^2)^{3/2}\, du = \dfrac{(u^2 + a^2)^{5/2}}{5}$

191. $\int u^2(u^2 + a^2)^{3/2}\, du = \dfrac{u(u^2 + a^2)^{5/2}}{6} - \dfrac{a^2 u(u^2 + a^2)^{3/2}}{24} - \dfrac{a^4 u\sqrt{u^2 + a^2}}{16} - \dfrac{a^6}{16}\ln\left| u + \sqrt{u^2 + a^2} \right|$

192. $\int u^3(u^2 + a^2)^{3/2}\, du = \dfrac{(u^2 + a^2)^{7/2}}{7} - \dfrac{a^2(u^2 + a^2)^{5/2}}{5}$

193. $\int \dfrac{(u^2 + a^2)^{3/2}}{u}\, du = \dfrac{(u^2 + a^2)^{3/2}}{3} + a^2\sqrt{u^2 + a^2} - a^3 \ln\left| \dfrac{a + \sqrt{u^2 + a^2}}{u} \right|$

194. $\int \dfrac{(u^2 + a^2)^{3/2}}{u^2}\, du = \dfrac{(u^2 + a^2)^{3/2}}{u} + \dfrac{3u\sqrt{u^2 + a^2}}{2} + \dfrac{3}{2}a^2 \ln\left| u^2 + a^2 \right|$

195. $\int \dfrac{(u^2 + a^2)^{3/2}}{u^3}\, du = \dfrac{(u^2 + a^2)^{3/2}}{2u^2} + \dfrac{3}{2}\sqrt{u^2 + a^2} - \dfrac{3}{2}a \ln\left| \dfrac{a + \sqrt{u^2 + a^2}}{u} \right|$

INTEGRALS INVOLVING $\sqrt{u^2 - a^2}$

---

196. $\int \dfrac{du}{\sqrt{u^2 - a^2}} = \ln\left| u + \sqrt{u^2 - a^2} \right|$

197. $\int \dfrac{u\, du}{\sqrt{u^2 - a^2}} = \sqrt{u^2 - a^2}$

198. $\int \dfrac{u^2\, du}{\sqrt{u^2 - a^2}} = \dfrac{u\sqrt{u^2 - a^2}}{2} + \dfrac{a^2}{2}\ln\left| u + \sqrt{u^2 - a^2} \right|$

199. $\int \dfrac{u^3\, du}{\sqrt{u^2 - a^2}} = \dfrac{(u^2 - a^2)^{3/2}}{3} + a^2\sqrt{u^2 - a^2}$

200. $\int \dfrac{du}{u\sqrt{u^2 - a^2}} = \dfrac{1}{a}\sec^{-1}\left| \dfrac{u}{a} \right|$

201. $\int \dfrac{du}{u^2\sqrt{u^2 - a^2}} = \dfrac{\sqrt{u^2 - a^2}}{a^2 u}$

202. $\int \dfrac{du}{u^3\sqrt{u^2 - a^2}} = \dfrac{\sqrt{u^2 - a^2}}{2a^2 u^2} + \dfrac{1}{2a^3}\sec^{-1}\left| \dfrac{u}{a} \right|$

203. $\int \sqrt{u^2 - a^2}\, du = \dfrac{u\sqrt{u^2 - a^2}}{2} - \dfrac{a^2}{2}\ln\left| u + \sqrt{u^2 - a^2} \right|$

204. $\int u\sqrt{u^2 - a^2}\, du = \dfrac{(u^2 - a^2)^{3/2}}{3}$

205. $\int u^2\sqrt{u^2-a^2}\,du = \dfrac{u(u^2-a^2)^{3/2}}{4} + \dfrac{a^2u\sqrt{u^2-a^2}}{8} - \dfrac{a^4}{8}\ln\left|u+\sqrt{u^2-a^2}\right|$

206. $\int u^3\sqrt{u^2-a^2}\,du = \dfrac{(u^2-a^2)^{5/2}}{5} + \dfrac{a^2(u^2-a^2)^{3/2}}{3}$

207. $\int \dfrac{\sqrt{u^2-a^2}}{u}\,du = \sqrt{u^2-a^2} - a\sec^{-1}\left|\dfrac{u}{a}\right|$

208. $\int \dfrac{\sqrt{u^2-a^2}}{u^2}\,du = -\dfrac{\sqrt{u^2-a^2}}{u} + \ln\left|u+\sqrt{u^2-a^2}\right|$

209. $\int \dfrac{\sqrt{u^2-a^2}}{u^3}\,du = -\dfrac{\sqrt{u^2-a^2}}{2u^2} + \dfrac{1}{2a}\sec^{-1}\left|\dfrac{u}{a}\right|$

210. $\int \dfrac{du}{(u^2-a^2)^{3/2}} = -\dfrac{u}{a^2\sqrt{u^2-a^2}}$

211. $\int \dfrac{u\,du}{(u^2-a^2)^{3/2}} = \dfrac{-1}{\sqrt{u^2-a^2}}$

212. $\int \dfrac{u^2\,du}{(u^2-a^2)^{3/2}} = -\dfrac{u}{\sqrt{u^2-a^2}} + \ln\left|u+\sqrt{u^2-a^2}\right|$

213. $\int \dfrac{u^3\,du}{(u^2-a^2)^{3/2}} = \sqrt{u^2-a^2} - \dfrac{a^2}{\sqrt{u^2-a^2}}$

214. $\int \dfrac{du}{u(u^2-a^2)^{3/2}} = \dfrac{-1}{a^2\sqrt{u^2-a^2}} - \dfrac{1}{a^3}\sec^{-1}\left|\dfrac{u}{a}\right|$

215. $\int \dfrac{du}{u^2(u^2-a^2)^{3/2}} = -\dfrac{\sqrt{u^2-a^2}}{a^4u} - \dfrac{u}{a^4\sqrt{u^2-a^2}}$

216. $\int \dfrac{du}{u^3(u^2-a^2)^{3/2}} = \dfrac{1}{2a^2u^2\sqrt{u^2-a^2}} - \dfrac{3}{2a^4\sqrt{u^2-a^2}} - \dfrac{3}{2a^5}\sec^{-1}\left|\dfrac{u}{a}\right|$

217. $\int (u^2-a^2)^{3/2}\,du = \dfrac{u(u^2-a^2)^{3/2}}{4} - \dfrac{3a^2u\sqrt{u^2-a^2}}{8} + \dfrac{3}{8}a^4\ln\left|u+\sqrt{u^2-a^2}\right|$

218. $\int u(u^2-a^2)^{3/2}\,du = \dfrac{(u^2+a^2)^{5/2}}{5}$

219. $\int u^2(u^2-a^2)^{3/2}\,du$

$= \dfrac{u(u^2-a^2)^{5/2}}{6} + \dfrac{a^2u(u^2-a^2)^{3/2}}{24} - \dfrac{a^4u\sqrt{u^2-a^2}}{16} + \dfrac{a^6}{16}\ln\left|u+\sqrt{u^2-a^2}\right|$

220. $\displaystyle\int u^3(u^2 - a^2)^{3/2}\, du = \frac{(u^2 - a^2)^{7/2}}{7} + \frac{a^2 u(u^2 - a^2)^{5/2}}{5}$

221. $\displaystyle\int \frac{(u^2 - a^2)^{3/2}}{u}\, du = \frac{(u^2 - a^2)^{3/2}}{3} - a^2\sqrt{u^2 - a^2} + a^3 \sec^{-1}\left|\frac{u}{a}\right|$

222. $\displaystyle\int \frac{(u^2 - a^2)^{3/2}}{u^2}\, du = \frac{(u^2 - a^2)^{3/2}}{u} + \frac{3u\sqrt{x^2 - a^2}}{2} - \frac{3}{2}a^2 \,\ln\left|u + \sqrt{u^2 - a^2}\right|$

223. $\displaystyle\int \frac{(u^2 - a^2)^{3/2}}{u^3}\, du = \frac{(u^2 - a^2)^{3/2}}{2u^2} + \frac{3\sqrt{x^2 - a^2}}{2} - \frac{3}{2}a \sec^{-1}\left|\frac{u}{a}\right|$

INTEGRALS INVOLVING $\sqrt{a^2 - u^2}$

---

224. $\displaystyle\int \frac{du}{\sqrt{a^2 - u^2}} = \sin^{-1}\frac{u}{a}$

225. $\displaystyle\int \frac{u\, du}{\sqrt{a^2 - u^2}} = -\sqrt{a^2 - u^2}$

226. $\displaystyle\int \frac{u^2\, du}{\sqrt{a^2 - u^2}} = -\frac{u\sqrt{a^2 - u^2}}{2} + \frac{a^2}{2}\sin^{-1}\frac{u}{a}$

227. $\displaystyle\int \frac{u^3\, du}{\sqrt{a^2 - u^2}} = \frac{(a^2 - u^2)^{3/2}}{3} - a^2\sqrt{a^2 - u^2}$

228. $\displaystyle\int \frac{du}{u\sqrt{a^2 - u^2}} = -\frac{1}{a}\ln\left|\frac{a + \sqrt{a^2 - u^2}}{u}\right|$

229. $\displaystyle\int \frac{du}{u^2\sqrt{a^2 - u^2}} = -\frac{\sqrt{a^2 - u^2}}{a^2 u}$

230. $\displaystyle\int \frac{du}{u^3\sqrt{a^2 - u^2}} = -\frac{\sqrt{a^2 - u^2}}{2a^2 u^2} - \frac{1}{2a^3}\ln\left|\frac{a + \sqrt{a^2 - u^2}}{u}\right|$

231. $\displaystyle\int \sqrt{a^2 - u^2}\, du = \frac{u\sqrt{a^2 - u^2}}{2} + \frac{a^2}{2}\sin^{-1}\frac{u}{a}$

232. $\displaystyle\int u\sqrt{a^2 - u^2}\, du = -\frac{(a^2 - u^2)^{3/2}}{3}$

233. $\displaystyle\int u^2\sqrt{a^2 - u^2}\, du = -\frac{u(a^2 - u^2)^{3/2}}{4} + \frac{a^2 u\sqrt{a^2 - u^2}}{8} + \frac{a^4}{8}\sin^{-1}\frac{u}{a}$

234. $\displaystyle\int u^3\sqrt{a^2 - u^2}\, du = \frac{(a^2 - u^2)^{5/2}}{5} - \frac{a^2(a^2 - u^2)^{3/2}}{3}$

235. $\int \dfrac{\sqrt{a^2 - u^2}}{u} \, du = \sqrt{a^2 - u^2} \, - \, a \ln\left| \dfrac{a + \sqrt{a^2 - u^2}}{u} \right|$

236. $\int \dfrac{\sqrt{a^2 - u^2}}{u^2} \, du = -\dfrac{\sqrt{a^2 - u^2}}{u} \, - \, \sin^{-1}\dfrac{u}{a}$

237. $\int \dfrac{\sqrt{a^2 - u^2}}{u^3} \, du = -\dfrac{\sqrt{a^2 - u^2}}{2u^2} + \dfrac{1}{2a} \ln\left| \dfrac{a + \sqrt{a^2 - u^2}}{u} \right|$

238. $\int \dfrac{du}{(a^2 - u^2)^{3/2}} = \dfrac{u}{a^2 \sqrt{a^2 - u^2}}$

239. $\int \dfrac{u \, du}{(a^2 - u^2)^{3/2}} = \dfrac{1}{\sqrt{a^2 - u^2}}$

240. $\int \dfrac{u^2 \, du}{(a^2 - u^2)^{3/2}} = \dfrac{u}{\sqrt{a^2 - u^2}} \, - \, \sin^{-1}\dfrac{u}{a}$

241. $\int \dfrac{u^3 \, du}{(a^2 - u^2)^{3/2}} = \sqrt{a^2 - u^2} + \dfrac{a^2}{\sqrt{a^2 - u^2}}$

242. $\int \dfrac{du}{u(a^2 - u^2)^{3/2}} = \dfrac{1}{a^2 \sqrt{a^2 - u^2}} \, - \, \dfrac{1}{a^3} \ln\left| \dfrac{a + \sqrt{a^2 - u^2}}{u} \right|$

243. $\int \dfrac{du}{u^2(a^2 - u^2)^{3/2}} = -\dfrac{\sqrt{a^2 - u^2}}{a^4 u} + \dfrac{u}{a^4 \sqrt{a^2 - u^2}}$

244. $\int \dfrac{du}{u^3(a^2 - u^2)^{3/2}} = \dfrac{-1}{2a^2 u^2 \sqrt{a^2 - u^2}} + \dfrac{3}{2a^4 \sqrt{a^2 - u^2}} \, - \, \dfrac{3}{2a^5} \ln\left| \dfrac{a + \sqrt{a^2 - u^2}}{u} \right|$

245. $\int (a^2 - u^2)^{3/2} \, du = \dfrac{u(a^2 - u^2)^{3/2}}{4} + \dfrac{3a^2 u \sqrt{a^2 - u^2}}{8} + \dfrac{3}{8}a^4 \sin^{-1}\dfrac{u}{a}$

246. $\int u(a^2 - u^2)^{3/2} \, du = -\dfrac{(a^2 - u^2)^{5/2}}{5}$

247. $\int u^2(a^2 - u^2)^{3/2} \, du = -\dfrac{u(a^2 - u^2)^{5/2}}{6} + \dfrac{a^2 u(a^2 - u^2)^{3/2}}{24} + \dfrac{a^4 u \sqrt{a^2 - u^2}}{16} + \dfrac{a^6}{16} \sin^{-1}\dfrac{u}{a}$

248. $\int u^3(a^2 - u^2)^{3/2} \, du = \dfrac{(a^2 - u^2)^{7/2}}{7} - \dfrac{a^2(a^2 - u^2)^{5/2}}{5}$

249. $\int \dfrac{(a^2 - u^2)^{3/2}}{u} \, du = \dfrac{(a^2 - u^2)^{3/2}}{3} + a^2 \sqrt{a^2 - u^2} \, - \, a^3 \ln\left| \dfrac{a + \sqrt{a^2 - u^2}}{u} \right|$

250. $\int \dfrac{(a^2 - u^2)^{3/2}}{u^2} \, du = -\dfrac{(a^2 - u^2)^{3/2}}{u} - \dfrac{3u\sqrt{a^2 - x^2}}{2} - \dfrac{3}{2}a^2 \sin^{-1}\dfrac{u}{a}$

251. $\int \dfrac{(a^2 - u^2)^{3/2}}{u^3} \, du = -\dfrac{(a^2 - u^2)^{3/2}}{2u^2} - \dfrac{3\sqrt{a^2 - x^2}}{2} + \dfrac{3}{2}a \ln\left| \dfrac{a + \sqrt{a^2 - u^2}}{u} \right|$

INTEGRALS INVOLVING $\sqrt{au^2 + bu + c}$

252. $\displaystyle\int \sqrt{au^2 + bu + c}\; du = \frac{(2au + b)\sqrt{au^2 + bu + c}}{4a} + \frac{4ac - b^2}{8a}\int \frac{du}{\sqrt{au^2 + bu + c}}$

253. $\displaystyle\int u\sqrt{au^2 + bu + c}\; du$

$\qquad = \dfrac{(au^2 + bu + c)^{3/2}}{3a} - \dfrac{b(2ac + b)}{8a^2}\sqrt{au^2 + bx + c} - \dfrac{b(4ac - b^2)}{16a^2}\displaystyle\int \frac{du}{\sqrt{au^2 + bu + c}}$

254. $\displaystyle\int u^2\sqrt{au^2 + bu + c}\; du = \frac{6au - 5b}{24a^2}(au^2 + bx + c)^{3/2} + \frac{5b^2 - 4ac}{16a^2}\int \sqrt{au^2 + bu + c}\; du$

255. $\displaystyle\int (au^2 + bu + c)^{n+1/2}du = \frac{(2au + b)(au^2 + bu + c)^{n+1/2}}{4a(n + 1)} + \frac{(2n + 1)(4ac - b^2)}{8a(n + 1)}\int (au^2 + bu + c)^{n - 1/2}du$

256. $\displaystyle\int u(au^2 + bu + c)^{n+1/2}du = \frac{(au^2 + bu + c)^{n+3/2}}{a(2n + 3)} - \frac{b}{2a}\int (au^2 + bu + c)^{n+1/2}\; du$

257. $\displaystyle\int \frac{du}{\sqrt{au^2 + bu + c}} = \begin{cases} \dfrac{1}{\sqrt{a}}\ln\left|2\sqrt{a}\;\sqrt{au^2 + bu + c} + 2au + b)\right| \\[3mm] -\dfrac{1}{\sqrt{-a}}\sin^{-1}\left(\dfrac{2au + b}{\sqrt{4ac - b^2}}\right) \text{ or } \dfrac{1}{\sqrt{a}}\sinh^{-1}\left(\dfrac{2au + b}{\sqrt{4ac - b^2}}\right) \end{cases}$

258. $\displaystyle\int \frac{u\; du}{\sqrt{au^2 + bu + c}} = \frac{\sqrt{au^2 + bu + c}}{a} - \frac{b}{2a}\int \frac{du}{\sqrt{au^2 + bu + c}}$

259. $\displaystyle\int \frac{u^2\; du}{\sqrt{au^2 + bu + c}} = \frac{2au - 3b}{4a^2}\sqrt{au^2 + bu + c} + \frac{3b^2 - 4ac}{8a^2}\int \frac{du}{\sqrt{au^2 + bu + c}}$

260. $\displaystyle\int \frac{du}{u\sqrt{au^2 + bu + c}} = \begin{cases} -\dfrac{1}{\sqrt{c}}\ln\left|\dfrac{2\sqrt{c}\sqrt{au^2 + bu + c} + bu + 2c}{u}\right| \\[3mm] \dfrac{1}{\sqrt{-c}}\sin^{-1}\left(\dfrac{bu + 2c}{|u|\sqrt{b^2 - 4ac}}\right) \text{ or } -\dfrac{1}{\sqrt{c}}\sinh^{-1}\left(\dfrac{bu + 2c}{|u|\sqrt{4ac - b^2}}\right) \end{cases}$

261. $\displaystyle\int \frac{du}{u^2\sqrt{au^2 + bu + c}} = -\frac{\sqrt{au^2 + bu + c}}{cu} - \frac{b}{2c}\int \frac{du}{u\sqrt{au^2 + bu + c}}$

262. $\displaystyle\int \frac{\sqrt{au^2 + bu + c}}{u}\; du = \sqrt{au^2 + bu + c} + \frac{b}{2}\int \frac{du}{\sqrt{au^2 + bu + c}} + c\int \frac{du}{u\sqrt{au^2 + bu + c}}$

263. $\displaystyle\int \frac{\sqrt{au^2 + bu + c}}{u^2}\; du = -\frac{\sqrt{au^2 + bu + c}}{u} + a\int \frac{du}{\sqrt{au^2 + bu + c}} + \frac{b}{2}\int \frac{du}{u\sqrt{au^2 + bu + c}}$

264. $\displaystyle\int \frac{du}{(au^2 + bu + c)^{3/2}} = \frac{2(2au + b)}{(4ac - b^2)\sqrt{au^2 + bu + c}}$

265. $\displaystyle\int \frac{u\,du}{(au^2 + bu + c)^{3/2}} = \frac{2(bu + 2c)}{(b^2 - 4ac)\sqrt{au^2 + bu + c}}$

266. $\displaystyle\int \frac{u^2\,du}{(au^2 + bu + c)^{3/2}} = \frac{(2b^2 - 4ac)u + 2bc}{a(4ac - b^2)\sqrt{au^2 + bu + c}} + \frac{1}{a}\int \frac{du}{\sqrt{au^2 + bu + c}}$

267. $\displaystyle\int \frac{du}{u(au^2 + bu + c)^{3/2}} = \frac{1}{c\sqrt{au^2 + bu + c}} + \frac{1}{c}\int \frac{du}{u\sqrt{au^2 + bu + c}} - \frac{b}{2c}\int \frac{du}{(au^2 + bu + c)^{3/2}}$

268. $\displaystyle\int \frac{du}{u^2(au^2+bu+c)^{3/2}}$

$\displaystyle = -\frac{au^2 + 2bu + c}{c^2 u\sqrt{au^2+bu+c}} + \frac{b^2 - 2ac}{2c^2}\int \frac{du}{(au^2 + bu + c)^{3/2}} - \frac{3b}{2c^2}\int \frac{du}{u\sqrt{au^2 + bu + c}}$

269. $\displaystyle\int \frac{du}{(au^2 + bu + c)^{n+1/2}}$

$\displaystyle = \frac{2(2au + b)}{(2n - 1)(4ac - b^2)(au^2 + bu + c)^{n - 1/2}} + \frac{8a(n - 1)}{(2n - 1)(4ac - b^2)}\int \frac{du}{(au^2 + bu + c)^{n - 1/2}}$

270. $\displaystyle\int \frac{du}{u(au^2 + bu + c)^{n + 1/2}}$

$\displaystyle = \frac{1}{(2n - 1)c(au^2 + bu + c)^{n - 1/2}} + \frac{1}{c}\int \frac{du}{u(au^2 + bu + c)^{n - 1/2}} - \frac{b}{2c}\int \frac{du}{(au^2 + bu + c)^{n+1/2}}$

INTEGRALS INVOLVING $u^3 + a^3$

---

271. $\displaystyle\int \frac{du}{u^3 + a^3} = \frac{1}{6a^2}\ln\left|\frac{(u + a)^2}{u^2 - au + a^2}\right| + \frac{1}{a^2\sqrt{3}}\tan^{-1}\frac{2u - a}{a\sqrt{3}}$

272. $\displaystyle\int \frac{u\,du}{u^3 + a^3} = \frac{1}{6a}\ln\left|\frac{u^2 - au + a^2}{(u + a)^2}\right| + \frac{1}{a\sqrt{3}}\tan^{-1}\frac{2u - a}{a\sqrt{3}}$

273. $\displaystyle\int \frac{u^2\,du}{u^3 + a^3} = \frac{1}{3}\ln\left|u^3 + a^3\right|$

274. $\displaystyle\int \frac{du}{u(u^3 + a^3)} = \frac{1}{3a^3}\ln\left|\frac{u^3}{u^3 + a^3}\right|$

275. $\displaystyle\int \frac{du}{u^2(u^3 + a^3)} = -\frac{1}{a^3 u} - \frac{1}{6a^4} \ln\left|\frac{u^2 - au + a^2}{(u+a)^2}\right| - \frac{1}{a^4\sqrt{3}} \tan^{-1}\frac{2u - a}{a\sqrt{3}}$

276. $\displaystyle\int \frac{du}{(u^3 + a^3)} = \frac{u}{3a^3(u^3 + a^3)} + \frac{1}{9a^5} \ln\left|\frac{(u+a)^2}{u^2 - au + a^2}\right| + \frac{2}{3a^5\sqrt{3}} \tan^{-1}\frac{2u - a}{a\sqrt{3}}$

277. $\displaystyle\int \frac{u\,du}{(u^3 + a^3)^2} = \frac{u^2}{3a^3(u^3 + a^3)} + \frac{1}{18a^4} \ln\left|\frac{u^2 - au + a^2}{(u+a)^2}\right| + \frac{1}{3a^4\sqrt{3}} \tan^{-1}\frac{2u - a}{a\sqrt{3}}$

278. $\displaystyle\int \frac{u^2\,du}{(u^3 + a^3)^2} = -\frac{1}{3(u^3 + a^3)}$

279. $\displaystyle\int \frac{du}{u(u^3 + a^3)^2} = \frac{1}{3a^3(u^3 + a^3)} + \frac{1}{3a^6} \ln\left|\frac{u^3}{u^3 + a^3}\right|$

280. $\displaystyle\int \frac{du}{u^2(u^3 + a^3)^2} = -\frac{1}{a^6 u} - \frac{u^2}{3a^6(u^3 + a^3)} - \frac{4}{3a^6} \int \frac{u\,du}{u^3 + a^3}$

281. $\displaystyle\int \frac{u^m\,du}{u^3 + a^3} = \frac{u^{m-2}}{m - 2} - a^3 \int \frac{u^{m-3}\,du}{u^3 + a^3}$

282. $\displaystyle\int \frac{du}{u^n(u^3 + a^3)} = \frac{-1}{a^3(n - 1)u^{n-1}} - \frac{1}{a^3} \int \frac{du}{u^{n-3}(u^3 + a^3)}$

INTEGRALS INVOLVING $u^4 \pm a^4$

283. $\displaystyle\int \frac{du}{u^4 + a^4} = \frac{1}{4a^3\sqrt{2}} \ln\left|\frac{u^2 + ax\sqrt{2} + a^2}{u^2 - au\sqrt{2} + a^2}\right| - \frac{1}{2a^3\sqrt{2}} \tan^{-1}\frac{au\sqrt{2}}{u^2 - a^2}$

284. $\displaystyle\int \frac{u\,du}{u^4 + a^4} = \frac{1}{2a^2} \tan^{-1}\frac{u^2}{a^2}$

285. $\displaystyle\int \frac{u^2\,du}{u^4 + a^4} = \frac{1}{4a\sqrt{2}} \ln\left|\frac{u^2 - ax\sqrt{2} + a^2}{u^2 + au\sqrt{2} + a^2}\right| - \frac{1}{2a\sqrt{2}} \tan^{-1}\frac{au\sqrt{2}}{u^2 - a^2}$

286. $\displaystyle\int \frac{u^3\,du}{u^4 + a^4} = \frac{1}{4} \ln(u^4 + a^4)$

287. $\displaystyle\int \frac{du}{u(u^4 + a^4)} = \frac{1}{4a^4} \ln\left(\frac{u^4}{u^4 + a^4}\right)$

288. $\displaystyle\int \frac{du}{u^2(u^4 + a^4)} = -\frac{1}{a^4 u} - \frac{1}{4a^5\sqrt{2}} \ln\left|\frac{u^2 - ax\sqrt{2} + a^2}{u^2 + au\sqrt{2} + a^2}\right| + \frac{1}{2a^5\sqrt{2}} \tan^{-1}\frac{au\sqrt{2}}{u^2 - a^2}$

289. $\displaystyle\int \frac{du}{u^3(u^4 + a^4)} = -\frac{1}{2a^4 u^2} - \frac{1}{2a^6} \tan^{-1}\frac{u^2}{a^2}$

290. $\int \dfrac{du}{u^4 - a^4} = \dfrac{1}{4a^3} \ln\left|\dfrac{u - a}{u + a}\right| - \dfrac{1}{2a^3} \tan^{-1}\dfrac{u}{a}$

291. $\int \dfrac{u\,du}{u^4 - a^4} = \dfrac{1}{4a^2} \ln\left|\dfrac{u^2 - a^2}{u^2 + a^2}\right|$

292. $\int \dfrac{u^2\,du}{u^4 - a^4} = \dfrac{1}{4a} \ln\left|\dfrac{u - a}{u + a}\right| + \dfrac{1}{2a} \tan^{-1}\dfrac{u}{a}$

293. $\int \dfrac{u^3\,du}{u^4 - a^4} = \dfrac{1}{4} \ln|u^4 - a^4|$

294. $\int \dfrac{du}{u(u^4 - a^4)} = \dfrac{1}{4a^4} \ln\left|\dfrac{u^4 - a^4}{u^4}\right|$

295. $\int \dfrac{du}{u^2(u^4 - a^4)} = \dfrac{1}{a^4 u} + \dfrac{1}{4a^5} \ln\left|\dfrac{u - a}{u + a}\right| + \dfrac{1}{2a^5} \tan^{-1}\dfrac{u}{a}$

296. $\int \dfrac{du}{u^3(u^4 - a^4)} = \dfrac{1}{2a^4 u^2} + \dfrac{1}{4a^6} \ln\left|\dfrac{u^2 - a^2}{u^2 + a^2}\right|$

INTEGRALS INVOLVING $u^n \pm a^n$

---

297. $\int \dfrac{du}{u(u^n - a^n)} = \dfrac{1}{na^n} \ln\left|\dfrac{u^n}{u^n + a^n}\right|$

298. $\int \dfrac{u^{n-1}\,du}{u^n + a^n} = \dfrac{1}{n} \ln|u^n + a^n|$

299. $\int \dfrac{u^m\,du}{(u^n + a^n)^r} = \int \dfrac{u^{m-n}\,du}{(u^n + a^n)^{r-1}} - a^n \int \dfrac{u^{m-n}\,du}{(u^n + a^n)^r}$

300. $\int \dfrac{du}{u^m(u^n + a^n)^r} = \dfrac{1}{a^n} \int \dfrac{du}{u^m(u^n + a^n)^{r-1}} - \dfrac{1}{a^n} \int \dfrac{du}{u^{m-n}(u^n + a^n)^r}$

301. $\int \dfrac{du}{u\sqrt{u^n + a^n}} = \dfrac{1}{n\sqrt{a^n}} \ln\left|\dfrac{\sqrt{u^n + a^n} - \sqrt{a^n}}{\sqrt{u^n + a^n} + \sqrt{a^n}}\right|$

302. $\int \dfrac{du}{u(u^n - a^n)} = \dfrac{1}{na^n} \ln\left|\dfrac{u^n - a^n}{u^n}\right|$

303. $\int \dfrac{u^{n-1}\,du}{u^n + a^n} = \dfrac{1}{n} \ln|u^n + a^n|$

304. $\int \dfrac{u^m\,du}{(u^n + a^n)^r} = a^n \int \dfrac{u^{m-n}\,du}{(u^n + a^n)^r} + \int \dfrac{u^{m-n}\,du}{(u^n + a^n)^{r-1}}$

305. $\displaystyle\int \frac{du}{u^m(u^n - a^n)^r} = \frac{1}{a^n} \int \frac{du}{u^{m-n}(u^n - a^n)^r} - \frac{1}{a^n} \int \frac{du}{u^m(u^n - a^n)^{r-1}}$

306. $\displaystyle\int \frac{du}{u\sqrt{u^n - a^n}} = \frac{2}{n\sqrt{a^n}} \cos^{-1} \sqrt{\frac{a^n}{u^n}}$

307. $\displaystyle\int \frac{u^{p-1}\,du}{u^{2m} + a^{2m}} = \frac{1}{ma^{2m-p}} \sum_{k=0}^{m} \sin \frac{(2k-1)p\pi}{2m} \tan^{-1}\left(\frac{u + a\cos[(2k-1)\pi/2m]}{a\sin[(2k-2)\pi/2m]}\right)$

$\displaystyle - \frac{1}{2ma^{2m-p}} \sum_{k=0}^{m} \cos \frac{(2k-1)p\pi}{2m} \ln\left| u^2 + 2au\cos\frac{(2k-1)}{2m} + a^2 \right|$ where $0 < p \le 2m$.

308. $\displaystyle\int \frac{u^{p-1}\,du}{u^{2m} - a^{2m}} = \frac{1}{2ma^{2m-p}} \sum_{k=1}^{m-1} \cos \frac{kp\pi}{m}\left(u^2 - 2au\cos\frac{k\pi}{m} + a^2 \right)$

$\displaystyle - \frac{1}{ma^{2m-p}} \sum_{k=1}^{m-1} \sin \frac{kp\pi}{m} \tan^{-1}\left(\frac{u - a\cos(k\pi/m)}{a\sin(k\pi/m)}\right)$ where $0 < p \le 2m$.

309. $\displaystyle\int \frac{u^{p-1}\,du}{u^{2m+1} + a^{2m+1}}$

$\displaystyle = \frac{2(-1)^{p-1}}{(2m+1)a^{2m-p+1}} \sum_{k=1}^{m} \sin \frac{2kp\pi}{2m+1} \tan^{-1}\left(\frac{u + a\cos[(2k\pi/(2m+1)]}{a\sin[2k\pi/(2m+1)]}\right)$

$\displaystyle - \frac{(-1)^{p-1}}{(2m+1)a^{2m-p+1}} \sum_{k=1}^{m} \cos \frac{2kp\pi}{2m+1} \ln\left| u^2 + 2au\cos\frac{2k\pi}{2m+1} + a^2 \right| + \frac{(-1)^{p-1}\ln|u + a|}{(2m+1)a^{2m-p+1}}$

where $0 < p \le 2m+1$

310. $\displaystyle\int \frac{u^{p-1}\,du}{u^{2m+1} + a^{2m+1}}$

$\displaystyle = \frac{-2}{(2m+1)a^{2m-p+1}} \sum_{k=1}^{m} \sin \frac{2kp\pi}{2m+1} \tan^{-1}\left(\frac{u - a\cos[2k\pi/(2m+1)]}{a\sin[2k\pi/(2m+1)]}\right)$

$\displaystyle + \frac{1}{(2m+1)a^{2m-p+1}} \sum_{k=1}^{m} \cos \frac{2kp\pi}{2m+1} \ln\left| u^2 - 2au\cos\frac{2k\pi}{2m+1} + a^2 \right| + \frac{\ln|u - a|}{(2m+1)a^{2m-p+1}}$

where $0 < p \le 2m+1$

INTEGRALS INVOLVING cos *au*

311. $\displaystyle\int \cos au\,du = \frac{\sin au}{a}$

312. $\displaystyle\int u\cos au\,du = \frac{\cos au}{a^2} + \frac{u\sin au}{a}$

313. $\int u^2 \cos au \, du = \dfrac{2u}{a^2} \cos au + \left(\dfrac{u^2}{a} - \dfrac{2}{a^3} \right) \sin au$

314. $\int u^3 \cos au \, du = \left(\dfrac{3u^2}{a^2} - \dfrac{6}{a^4} \right) \cos au + \left(\dfrac{u^3}{a} - \dfrac{6u}{a^3} \right) \sin au$

315. $\int u^n \cos au \, du = \dfrac{u^n \sin au}{a} - \dfrac{n}{a} \int u^{n-1} \sin au \, du$

316. $\int u^n \cos au \, du = \dfrac{u^n \sin au}{a} + \dfrac{nu^{n-1}}{a^2} \cos au - \dfrac{n(n-1)}{a^2} \int u^{n-2} \cos au \, du$

317. $\int \cos^2 au \, du = \dfrac{u}{2} + \dfrac{\sin 2au}{4a}$

318. $\int u \cos^2 au \, du = \dfrac{u^2}{4} + \dfrac{u \sin 2au}{4a} + \dfrac{\cos 2au}{8a^2}$

319. $\int \cos^3 au \, du = \dfrac{\sin au}{a} - \dfrac{\sin^3 au}{3a}$

320. $\int \cos^4 au \, du = \dfrac{3u}{8} - \dfrac{\sin 2au}{4a} + \dfrac{\sin 4au}{32a}$

321. $\int \cos^n au \, du = \dfrac{\cos^{n-1} au \sin au}{an} + \dfrac{n-1}{n} \int \cos^{n-2} au \, du$

322. $\int \dfrac{\cos au}{u} \, du = \ln|u| - \dfrac{(au)^2}{2 \cdot 2!} + \dfrac{(au)^4}{4 \cdot 4!} - \dfrac{(au)^6}{6 \cdot 6!} + \cdots$

323. $\int \dfrac{\cos au}{u^2} \, du = -\dfrac{\cos au}{u} - a \int \dfrac{\sin au}{u} \, du$

324. $\int \dfrac{\cos au}{u^n} \, du = -\dfrac{\cos au}{(n-1)u^{n-1}} - \dfrac{a}{n-1} \int \dfrac{\sin au}{u^{n-1}} \, du$

325. $\int \dfrac{du}{\cos au} = \dfrac{1}{a} \ln|\sec au + \tan au| = \dfrac{1}{a} \ln \tan \left| \dfrac{\pi}{4} + \dfrac{au}{2} \right|$

326. $\int \dfrac{u \, du}{\cos au} = \dfrac{1}{a^2} \left\{ \dfrac{(au)^2}{2} + \dfrac{(au)^4}{8} + \dfrac{5(au)^6}{144} + \cdots + \dfrac{E_n(au)^{2n+2}}{(2n+2)(2n)!} + \cdots \right\}$

327. $\int \dfrac{du}{\cos^2 au} = \dfrac{\tan au}{a}$

328. $\int \dfrac{du}{\cos^3 au} = \dfrac{\sin au}{2a \cos^2 au} + \dfrac{1}{2a} \ln \left| \tan \left(\dfrac{\pi}{4} + \dfrac{au}{2} \right) \right|$

329. $\displaystyle\int \frac{du}{\cos^n au} = \frac{\sin au}{a(n-1)\cos^{n-1} au} + \frac{n-2}{n-1} \int \frac{du}{\cos^{n-2} au}$

330. $\displaystyle\int \frac{u\,du}{\cos^n au} = \frac{u \sin au}{a(n-1)\cos^{n-1} au} - \frac{1}{a^2(n-1)(n-2)\cos^{n-2} au} + \frac{n-2}{n-1} \int \frac{u\,du}{\cos^{n-2} au}$

331. $\displaystyle\int \cos pu \cos pu\,du = \frac{\sin(a-p)u}{2(a-p)} + \frac{\sin(a+`p)u}{2(a+p)}$

332. $\displaystyle\int \frac{du}{1-\cos au} = -\frac{1}{a}\cot \frac{au}{2}$

333. $\displaystyle\int \frac{u\,du}{1-\cos au} = -\frac{u}{a}\cot \frac{au}{2} + \frac{2}{a^2}\ln\left|\sin \frac{au}{2}\right|$

334. $\displaystyle\int \frac{du}{1+\cos au} = \frac{1}{a}\tan \frac{au}{2}$

335. $\displaystyle\int \frac{u\,du}{1+\cos au} = \frac{u}{a}\tan \frac{au}{2} + \frac{2}{a^2}\ln\left|\cos \frac{au}{2}\right|$

336. $\displaystyle\int \frac{du}{(1-\cos au)^2} = -\frac{1}{2a}\cot \frac{au}{2} - \frac{1}{6a}\cot^3 \frac{au}{2}$

337. $\displaystyle\int \frac{du}{(1+\cos au)^2} = \frac{1}{2a}\tan \frac{au}{2} + \frac{1}{6a}\tan^3 \frac{au}{2}$

338. $\displaystyle\int \frac{du}{p+q\cos au} = \begin{cases} \dfrac{2}{a\sqrt{p^2-q^2}}\tan^{-1}\sqrt{(p-q)/(p+q)}\tan\tfrac{1}{2}au \\[4mm] \dfrac{1}{a\sqrt{q^2-p^2}}\ln\left|\dfrac{\tan\frac{1}{2}au + \sqrt{(q+p)/(q-p)}}{\tan\frac{1}{2}au - \sqrt{(q+p)/(q-p)}}\right| \end{cases}$

339. $\displaystyle\int \frac{du}{(p+q\cos au)^2} = \frac{q\sin au}{a(q^2-p^2)(p+q\cos au)} - \frac{p}{q^2-p^2}\int \frac{du}{p+q\cos au}$

340. $\displaystyle\int \frac{du}{p^2+q^2\cos^2 au} = \frac{1}{ap\sqrt{p^2+q^2}}\tan^{-1}\frac{p\tan au}{\sqrt{p^2+q^2}}$

341. $\displaystyle\int \frac{du}{p^2-q^2\cos^2 au} = \begin{cases} \dfrac{1}{ap\sqrt{p^2-q^2}}\tan^{-1}\dfrac{p\tan au}{\sqrt{p^2-q^2}} \\[4mm] \dfrac{1}{2ap\sqrt{q^2-p^2}}\ln\left|\dfrac{p\tan au - \sqrt{q^2-p^2}}{p\tan au + \sqrt{q^2-p^2}}\right| \end{cases}$

INTEGRALS INVOLVING sin au

342. $\int \sin\ au\ du = -\dfrac{\cos\ au}{a}$

343. $\int u \sin\ au\ du = \dfrac{\sin\ au}{a^2} - \dfrac{u \cos\ au}{a}$

344. $\int u^2 \sin\ au\ du = \dfrac{2u}{a^2}\sin\ au + \left(\dfrac{2}{a^3} - \dfrac{u^2}{a}\right)\cos\ au$

345. $\int u^3 \sin\ au\ du = \left(\dfrac{3u^2}{a^2} - \dfrac{6}{a^4}\right)\sin\ au + \left(\dfrac{6u}{a^3} - \dfrac{u^3}{a}\right)\cos\ au$

346. $\int u^n \sin\ au\ du = -\dfrac{u^n \cos\ au}{a} + \dfrac{n}{a}\int u^{n-1}\cos\ au\ du$

347. $\int u^n \sin\ au\ du = -\dfrac{u^n \cos\ au}{a} + \dfrac{nu^{n-1}\sin\ au}{a^2} - \dfrac{n(n-1)}{a^2}\int u^{n-2}\sin\ au\ du$

348. $\int \sin^2\ au\ du = \dfrac{u}{2} - \dfrac{\sin 2\ au}{4a}$

349. $\int u \sin^2\ au\ du = \dfrac{u^2}{4} - \dfrac{u \sin 2au}{4a} - \dfrac{\cos 2au}{8a^2}$

350. $\int \sin^3\ au\ du = -\dfrac{\cos\ au}{a} + \dfrac{\cos^3\ au}{3a}$

351. $\int \sin^4\ au\ du = \dfrac{3u}{8} - \dfrac{\sin 2au}{4a} + \dfrac{\sin 4au}{32a}$

352. $\int \sin^n\ au\ du = -\dfrac{\sin^{n-1}\ au \cos\ au}{an} + \dfrac{n-1}{n}\int \sin^{n-2}\ au\ du$

353. $\int \dfrac{\sin\ au}{u}\ du = au - \dfrac{(au)^3}{3\cdot 3!} + \dfrac{(au)^5}{5\cdot 5!} - \cdots$

354. $\int \dfrac{\sin\ au}{u^2}\ du = -\dfrac{\sin\ au}{u} + a\int \dfrac{\cos\ au}{u}\ du$

355. $\int \dfrac{\sin\ au}{u^n}\ du = -\dfrac{\sin\ au}{(n-1)u^{n-1}} + \dfrac{a}{n-1}\int \dfrac{\cos\ au}{u^{n-1}}\ du$

356. $\int \dfrac{du}{\sin\ au} = \dfrac{1}{a}\ln|\csc\ au - \cot\ au| = \dfrac{1}{a}\ln\left|\tan\dfrac{au}{2}\right|$

357. $\displaystyle\int \frac{u\,du}{\sin\,au} = \frac{1}{a^2}\left\{au + \frac{(au)^3}{18} + \frac{7(au)^5}{1800} + \cdots + \frac{2(2^{2n-1}-1)B_n(au)^{2n+1}}{(2n+1)!} + \cdots\right\}$

358. $\displaystyle\int \frac{du}{\sin^2\,au} = -\frac{1}{a}\cot\,au$

359. $\displaystyle\int \frac{du}{\sin^3\,au} = -\frac{\cos\,au}{2a\sin^2\,au} + \frac{1}{2a}\ln\left|\tan\frac{au}{2}\right|$

360. $\displaystyle\int \frac{du}{\sin^n\,au} = \frac{-\cos\,au}{a(n-1)\sin^{n-1}au} + \frac{n-2}{n-1}\int\frac{du}{\sin^{n-2}au}$

361. $\displaystyle\int \frac{u\,du}{\sin^n\,au} = \frac{-u\cos\,au}{a(n-1)\sin^{n-1}au} - \frac{1}{a^2(n-1)(n-2)\sin^{n-2}au} + \frac{n-2}{n-1}\int\frac{u\,du}{\sin^{n-2}au}$

362. $\displaystyle\int \sin\,pu\,\sin\,qu\,du = \frac{\sin(p-q)u}{2(p-q)} - \frac{\sin(p+q)u}{2(p+q)}$

363. $\displaystyle\int \frac{du}{1-\sin\,au} = \frac{1}{a}\tan\left(\frac{\pi}{4}+\frac{au}{2}\right)$

364. $\displaystyle\int \frac{u\,du}{1-\sin\,au} = \frac{u}{a}\tan\left(\frac{\pi}{4}+\frac{au}{2}\right) + \frac{2}{a^2}\ln\left|\sin\left(\frac{\pi}{4}-\frac{au}{2}\right)\right|$

365. $\displaystyle\int \frac{du}{1+\sin\,au} = -\frac{1}{a}\tan\left(\frac{\pi}{4}-\frac{au}{2}\right)$

366. $\displaystyle\int \frac{u\,du}{1+\sin\,au} = -\frac{u}{a}\tan\left(\frac{\pi}{4}-\frac{au}{2}\right) + \frac{2}{a^2}\ln\left|\sin\left(\frac{\pi}{4}+\frac{au}{2}\right)\right|$

367. $\displaystyle\int \frac{du}{(1-\sin\,au)^2} = \frac{1}{2a}\tan\left(\frac{\pi}{4}+\frac{au}{2}\right) + \frac{1}{6a}\tan^3\left(\frac{\pi}{4}+\frac{au}{2}\right)$

368. $\displaystyle\int \frac{du}{(1+\sin\,au)^2} = \frac{1}{2a}\tan\left(\frac{\pi}{4}-\frac{au}{2}\right) - \frac{1}{6a}\tan^3\left(\frac{\pi}{4}-\frac{au}{2}\right)$

369. $\displaystyle\int \frac{du}{p+q\sin\,au} = \begin{cases}\dfrac{2}{a\sqrt{p^2-q^2}}\tan^{-1}\dfrac{p\tan\frac{1}{2}au + q}{\sqrt{p^2-q^2}} \\[4mm] \dfrac{1}{a\sqrt{q^2-p^2}}\ln\left|\dfrac{p\tan\frac{1}{2}au + q - \sqrt{q^2-p^2}}{p\tan\frac{1}{2}au + q + \sqrt{q^2-p^2}}\right|\end{cases}$

370. $\displaystyle\int \frac{du}{(p+q\sin\,au)^2} = \frac{q\cos\,au}{a(p^2-q^2)(p+q\sin\,au)} + \frac{p}{p^2-q^2}\int\frac{du}{p+q\sin\,au}$

371. $\displaystyle\int \frac{du}{p^2+q^2\sin^2\,au} = \frac{1}{ap\sqrt{p^2+q^2}}\tan^{-1}\frac{\sqrt{p^2+q^2}\,\tan\,au}{p}$

372. $\displaystyle\int \frac{du}{p^2 - q^2 \sin^2 au} = \begin{cases} \dfrac{1}{ap\sqrt{p^2 - q^2}} \tan^{-1} \dfrac{\sqrt{p^2 - q^2}\, \tan\, au}{p} \\[4mm] \dfrac{1}{2ap\sqrt{q^2 - p^2}} \ln \left| \dfrac{\sqrt{q^2 - p^2}\, \tan\, au + p}{\sqrt{q^2 - p^2}\, \tan\, au - p} \right| \end{cases}$

INTEGRALS INVOLVING sin *au* and cos *au*

---

373. $\displaystyle\int \sin\, au \cos\, au \; du = \frac{\sin^2 au}{2a}$

374. $\displaystyle\int \sin\, pu \cos\, qu \; du = - \frac{\cos\,(p - q)u}{2(p - q)} - \frac{\cos(p + q)u}{2(p + q)}$

375. $\displaystyle\int \sin^n au \cos\, au \; du = \frac{\sin^{n+1} au}{(n + 1)a}$

376. $\displaystyle\int \cos^n au \sin\, au \; du = - \frac{\cos^{n+1} au}{(n + 1)a}$

377. $\displaystyle\int \sin^2 au \cos^2 au \; du = \frac{u}{8} - \frac{\sin 4au}{32a}$

378. $\displaystyle\int \frac{du}{\sin\, au \cos\, au} = \frac{1}{a} \ln| \tan\, au|$

379. $\displaystyle\int \frac{du}{\sin^2 au \cos\, au} = \frac{1}{a} \ln \left| \tan \left(\frac{\pi}{4} + \frac{au}{2} \right) \right| - \frac{1}{a \sin\, au}$

380. $\displaystyle\int \frac{du}{\sin\, au \cos^2 au} = \frac{1}{a} \ln \left| \tan \frac{au}{2} \right| + \frac{1}{a \cos\, au}$

381. $\displaystyle\int \frac{du}{\sin^2 au \cos^2 au} = - \frac{2 \cot 2\, au}{a}$

382. $\displaystyle\int \frac{\sin^2 au}{\cos\, au} \; du = - \frac{\sin\, au}{a} + \frac{1}{a} \ln \left| \tan \left(\frac{au}{2} + \frac{\pi}{4} \right) \right|$

383. $\displaystyle\int \frac{\cos^2 au}{\sin\, au} \; du = \frac{\cos\, au}{a} + \frac{1}{a} \ln \left| \tan \frac{au}{2} \right|$

384. $\displaystyle\int \frac{du}{\cos\, au(1 \pm \sin\, au)} = \mp \frac{1}{2a(1 \pm \sin\, au)} + \frac{1}{2a} \ln \left| \tan \left(\frac{au}{2} + \frac{\pi}{4} \right) \right|$

385. $\displaystyle\int \frac{du}{\sin au(1 \pm \cos au)} = \pm\frac{1}{2a(1 \pm \cos au)} + \frac{1}{2a}\ln\left|\tan\frac{au}{2}\right|$

386. $\displaystyle\int \frac{du}{\sin au \pm \cos au} = \frac{1}{a\sqrt{2}}\ln\left|\tan\left(\frac{au}{2} \pm \frac{\pi}{8}\right)\right|$

387. $\displaystyle\int \frac{\sin au\,du}{\sin au \pm \cos au} = \frac{u}{2} \mp \frac{1}{2a}\ln|\sin au \pm \cos au|$

388. $\displaystyle\int \frac{\cos au\,du}{\sin au \pm \cos au} = \mp\frac{u}{2} + \frac{1}{2a}\ln|\sin au \pm \cos au|$

389. $\displaystyle\int \frac{\sin au\,du}{p + q\cos au} = -\frac{1}{aq}\ln|p + q\cos au|$

390. $\displaystyle\int \frac{\cos au\,du}{p + q\sin au} = \frac{1}{aq}\ln|p + q\sin au|$

391. $\displaystyle\int \frac{\sin au\,du}{(p + q\cos au)^n} = \frac{1}{aq(n-1)(p + q\cos au)^{n-1}}$

392. $\displaystyle\int \frac{\cos au\,du}{(p + q\sin au)^n} = \frac{-1}{aq(n-1)(p + q\sin au)^{n-1}}$

393. $\displaystyle\int \frac{du}{p\sin au + q\cos au} = \frac{1}{a\sqrt{p^2 + q^2}}\ln\left|\tan\left(\frac{au + \tan^{-1}(q/p)}{2}\right)\right|$

394. $\displaystyle\int \frac{du}{p\sin au + q\cos au + r} = \begin{cases} \dfrac{2}{a\sqrt{r^2 - p^2 - q^2}}\tan^{-1}\left(\dfrac{p + (r-q)\tan(au/2)}{\sqrt{r^2 - p^2 - q^2}}\right) \\[3mm] \dfrac{1}{a\sqrt{p^2 + q^2 - r^2}}\ln\left|\dfrac{p - \sqrt{p^2 + q^2 - r^2} + (r-q)\tan(au/2)}{p + \sqrt{p^2 + q^2 - r^2} + (r-q)\tan(au/2)}\right| \end{cases}$

395. $\displaystyle\int \frac{du}{p\sin au + q(1 + \cos au)} = \frac{1}{ap}\ln\left|q + p\tan\frac{au}{2}\right|$

396. $\displaystyle\int \frac{du}{p\sin au + q\cos au \pm \sqrt{p^2 + q^2}} = \frac{-1}{a\sqrt{p^2 + q^2}}\tan\left(\frac{\pi}{4} \mp \frac{au + \tan^{-1}(q/p)}{2}\right)$

397. $\displaystyle\int \frac{du}{p^2\sin^2 au + q^2\cos^2 au} = \frac{1}{apq}\tan^{-1}\left(\frac{p\tan au}{q}\right)$

398. $\displaystyle\int \frac{du}{p^2\sin^2 au - q^2\cos^2 au} = \frac{1}{2apq}\tan\left(\frac{p\tan au - q}{p\tan au + q}\right)$

399. $\displaystyle\int \sin^m au \cos^n au\, du = \begin{cases} -\dfrac{\sin^{m-1} au \cos^{n+1} au}{a(m+n)} + \dfrac{m-1}{m+n} \displaystyle\int \sin^{m-2} au \cos^n au\, du \\[2ex] \dfrac{\sin^{m+1} au \cos^{n-1} au}{a(m+n)} + \dfrac{n-1}{m+n} \displaystyle\int \sin^m au \cos^{n-2} au\, du \end{cases}$

400. $\displaystyle\int \dfrac{\sin^m au}{\cos^n au}\, du = \begin{cases} \dfrac{\sin^{m-1} au}{a(n-1)\cos^{n-1} au} - \dfrac{m-1}{n-1} \displaystyle\int \dfrac{\sin^{m-2} au}{\cos^n au}\, du \\[2ex] \dfrac{\sin^{m+1} au}{a(n-1)\cos^{n-1} au} - \dfrac{m-n+2}{n-1} \displaystyle\int \dfrac{\sin^m au}{\cos^{n-2} au}\, du \\[2ex] \dfrac{-\sin^{m-1} au}{a(m-n)\cos^{n-1} au} + \dfrac{m-1}{n-1} \displaystyle\int \dfrac{\sin^{m-2} au}{\cos^n au}\, du \end{cases}$

401. $\displaystyle\int \dfrac{\cos^m au}{\sin^n au}\, du = \begin{cases} \dfrac{-\cos^{m-1} au}{a(n-1)\sin^{n-1} au} - \dfrac{m-1}{n-1} \displaystyle\int \dfrac{\cos^{m-2} au}{\sin^{n-2} au}\, du \\[2ex] \dfrac{-\cos^{m+1} au}{a(n-1)\sin^{n-1} au} - \dfrac{m-n+2}{n-1} \displaystyle\int \dfrac{\cos^m au}{\sin^{n-2} au}\, du \\[2ex] \dfrac{\cos^{m-1} au}{a(m-n)\sin^{n-1} au} + \dfrac{m-1}{n-1} \displaystyle\int \dfrac{\cos^{m-2} au}{\sin^n au}\, du \end{cases}$

402. $\displaystyle\int \dfrac{du}{\sin^m au \cos^n au} = \begin{cases} \dfrac{1}{a(n-1)\sin^{m-1} au \cos^{n-1} au} + \dfrac{m+n-2}{n-1} \displaystyle\int \dfrac{du}{\sin^m au \cos^{n-2} au} \\[2ex] \dfrac{1}{a(m-1)\sin^{m-1} au \cos^{n-1} au} + \dfrac{m+n-2}{m-1} \displaystyle\int \dfrac{du}{\sin^{m-2} au \cos^n au} \end{cases}$

INTEGRALS INVOLVING tan *au*

403. $\displaystyle\int \tan au\, du = -\frac{1}{a}\ln|\cos au| = \frac{1}{a}\ln|\sec au|$

404. $\displaystyle\int \tan^2 au\, du = \frac{\tan au}{a} - u$

405. $\displaystyle\int \tan^3 au\, du = \frac{\tan^2 au}{2a} + \frac{1}{a}\ln|\cos au|$

406. $\displaystyle\int \tan^n au\, du = \frac{\tan^{n-1} au}{(n-1)a} - \int \tan^{n-2} au\, du$

407. $\displaystyle\int \tan^n au \sec^2 au\, du = \frac{\tan^{n+1} au}{(n+1)a}$

408. $\displaystyle\int \frac{\sec^2 u}{\tan au}\, du = \frac{1}{a} \ln|\tan au|$

409. $\displaystyle\int \frac{du}{\tan au} = \frac{1}{a} \ln|\sin au|$

410. $\displaystyle\int u \tan au\, du = \frac{1}{a^2}\left\{\frac{(au)}{3} + \frac{(au)^5}{15} + \frac{2(au)^7}{105} + \cdots + \frac{2^{2n}(2^{2n}-1)B_n(au)^{2n+1}}{(2n+1)!} + \cdots\right\}$

411. $\displaystyle\int \frac{\tan au}{u}\, du = au + \frac{(au)^3}{9} + \frac{2(au)^5}{75} + \cdots + \frac{2^{2n}(2^{2n}-1)B_n(au)^{2n-1}}{(2n-1)(2n)!} + \cdots$

412. $\displaystyle\int u \tan^2 au\, du = \frac{u \tan au}{a} + \frac{1}{a^2} \ln|\cos au| - \frac{u^2}{2}$

413. $\displaystyle\int \frac{du}{p + q \tan au} = \frac{pu}{p^2+q^2} + \frac{q}{a(p^2+q^2)} \ln|q \sin au + p \cos au|$

INTEGRALS INVOLVING cot *au*

414. $\displaystyle\int \cot au\, du = \frac{1}{a} \ln|\sin au| = -\frac{1}{a}|\csc u|$

415. $\displaystyle\int \cot^2 au\, du = -\frac{\cot au}{a} - u$

416. $\displaystyle\int \cot^3 au\, du = -\frac{\cot^2 au}{2a} - \frac{1}{a} \ln|\sin au|$

417. $\displaystyle\int \cot^n au\, du = \frac{\cot^{n-1} au}{(n-1)a} - \int \cot^{n-2} au\, du$

418. $\displaystyle\int \cot^n au \csc^2 au\, du = -\frac{\cot^{n+1}}{(n+1)a}$

419. $\displaystyle\int \frac{\csc^2 au}{\cot au}\, du = -\frac{1}{a} \ln|\cot au|$

420. $\displaystyle\int \frac{du}{\cot au} = -\frac{1}{a} \ln|\cos au|$

421. $\displaystyle\int u \cot au\, du = \frac{1}{a^2}\left\{au - \frac{(au)^3}{9} - \frac{(au)^5}{225} - \cdots - \frac{2^{2n}B_n(au)^{2n+1}}{(2n+1)!} - \cdots\right\}$

422. $\displaystyle\int \frac{\cot au}{u}\, du = -\frac{1}{au} - \frac{au}{3} - \frac{(au)^3}{135} - \cdots - \frac{2^{2n}B_n(au)^{2n-1}}{(2n-1)(2n)!} - \cdots$

423. $\displaystyle\int u \cot^2 au\, du = \frac{u \cot au}{a} + \frac{1}{a^2} \ln|\sin au| - \frac{u^2}{2}$

424. $\displaystyle\int \frac{du}{p + q \cot au} = \frac{pu}{p^2+q^2} - \frac{q}{a(p^2+q^2)} \ln|p \sin au + q \cos au|$

INTEGRALS INVOLVING sec *au*

425. $\int \sec au \, du = \frac{1}{a} \ln|\sec au + \tan au| = \frac{1}{a} \ln\left|\tan\left(\frac{au}{2} + \frac{\pi}{4}\right)\right|$

426. $\int \sec^2 au \, du = \frac{\tan au}{a}$

427. $\int \sec^3 au \, du = \frac{\sec au \tan au}{2a} + \frac{1}{2a} \ln|\sec au + \tan au|$

428. $\int \sec^n au \, du = \frac{\sec^{n-2} au \tan au}{a(n-1)} + \frac{n-2}{n-1} \int \sec^{n-2} au \, du$

429. $\int \sec^n au \tan au \, du = \frac{\sec^n au}{na}$

430. $\int \frac{du}{\sec au} = \frac{\sin au}{a}$

431. $\int u \sec au \, du = \frac{1}{a^2}\left\{ \frac{(au)^2}{2} + \frac{(au)^4}{8} + \frac{5(au)^6}{144} + \cdots + \frac{E_n(au)^{2n+2}}{(2n+2)(2n)!} + \cdots \right\}$

432. $\int \frac{\sec au}{u} \, du = \ln|u| + \frac{(au)^2}{4} + \frac{5(au)^4}{96} + \frac{61(au)^6}{4320} + \cdots + \frac{E_n(au)^{2n}}{2n(2n)!} + \cdots$

433. $\int u \sec^2 au \, du = \frac{u}{a} \tan au + \frac{1}{a^2} \ln|\cos au|$

434. $\int \frac{du}{q + p \sec au} = \frac{u}{q} - \frac{p}{q} \int \frac{du}{p + q \cos au}$

INTEGRALS INVOLVING csc *au*

435. $\int \csc au \, du = \frac{1}{a} \ln|\csc au - \cot au| = \frac{1}{a} \ln\left|\tan \frac{au}{2}\right|$

436. $\int \csc^2 au \, du = -\frac{\cot au}{a}$

437. $\int \csc^3 au \, du = -\frac{\csc au \cot au}{2a} + \frac{1}{2a} \ln\left|\tan \frac{au}{2}\right|$

438. $\int \csc^n au \, du = \frac{\csc^{n-2} au \cot au}{a(n-1)} + \frac{n-2}{n-1} \int \csc^{n-2} au \, du$

439. $\displaystyle\int \csc^n au \cot u \ du = -\frac{\csc^n au}{na}$

440. $\displaystyle\int \frac{du}{\csc au} = -\frac{\cos au}{a}$

441. $\displaystyle\int u \csc au \ du = \frac{1}{a^2}\left\{ au + \frac{(au)^3}{18} + \frac{7(au)^5}{1800} + \cdots + \frac{2(2^{2n-1}-1)B_n(au)^{2n+1}}{(2n+1)!} + \cdots \right\}$

442. $\displaystyle\int \frac{\csc au}{u} \ du = -\frac{1}{au} + \frac{au}{6} + \frac{7(au)^3}{1080} + \cdots + \frac{2(2^{2n-1}-1)B_n(au)^{2n-1}}{(2n-1)(2n)!} + \cdots$

443. $\displaystyle\int u \csc^2 au \ du = -\frac{u \cot au}{a} + \frac{1}{a^2}\ln|\sin au|$

444. $\displaystyle\int \frac{du}{q + p \csc au} = \frac{u}{q} - \frac{p}{q}\int \frac{du}{p + q \sin au}$

INTEGRALS INVOLVING INVERSE TRIGONOMETRIC FUNCTIONS

445. $\displaystyle\int \cos^{-1}\frac{u}{a} \ du = u \cos^{-1}\frac{u}{a} - \sqrt{a^2 - u^2}$

446. $\displaystyle\int u \cos^{-1}\frac{u}{a} \ du = \left(\frac{u^2}{2} - \frac{a^2}{4}\right)\cos^{-1}\frac{u}{a} - \frac{u\sqrt{a^2 - u^2}}{4}$

447. $\displaystyle\int u^2 \cos^{-1}\frac{u}{a} \ du = \frac{u^3}{3}\cos^{-1}\frac{u}{a} - \frac{(u^2 + 2a^2)\sqrt{a^2 - u^2}}{9}$

448. $\displaystyle\int \frac{\cos^{-1}(u/a)}{u} \ du = \frac{\pi}{2}\ln|u| - \int \frac{\sin^{-1}(u/a)}{u} \ du$

449. $\displaystyle\int \frac{\cos^{-1}(u/a)}{u^2} \ du = -\frac{\cos^{-1}(u/a)}{u} + \frac{1}{a}\ln\left|\frac{a + \sqrt{a^2 - u^2}}{u}\right|$

450. $\displaystyle\int \left(\cos^{-1}\frac{u}{a}\right)^2 \ du = u\left(\cos^{-1}\frac{u}{a}\right)^2 - 2u - 2\sqrt{a^2 - u^2}\cos^{-1}\frac{u}{a}$

451. $\displaystyle\int \sin^{-1}\frac{u}{a} \ du = u \sin^{-1}\frac{u}{a} + \sqrt{a^2 - u^2}$

452. $\displaystyle\int u \sin^{-1}\frac{u}{a} \ du = \left(\frac{u^2}{2} - \frac{a^2}{4}\right)\sin^{-1}\frac{u}{a} + \frac{u\sqrt{a^2 - u^2}}{4}$

453. $\displaystyle\int u^2 \sin^{-1}\frac{u}{a} \ du = \frac{u^3}{3}\sin^{-1}\frac{u}{a} + \frac{(u^2 + 2a^2)\sqrt{a^2 - u^2}}{9}$

454. $\int \frac{\sin^{-1}(u/a)}{u} du = \frac{u}{a} + \frac{(u/a)^3}{2 \cdot 3 \cdot 3} + \frac{1 \cdot 3(u/a)^5}{2 \cdot 4 \cdot 5 \cdot 5} + \frac{1 \cdot 3 \cdot 5(u/a)^7}{2 \cdot 4 \cdot 6 \cdot 7 \cdot 7} + \cdots$

455. $\int \frac{\sin^{-1}(u/a)}{u^2} du = -\frac{\sin^{-1}(u/a)}{u} - \frac{1}{a} \ln\left|\frac{a + \sqrt{a^2 - u^2}}{u}\right|$

456. $\int \left(\sin^{-1}\frac{u}{a}\right)^2 du = u\left(\sin^{-1}\frac{u}{a}\right)^2 - 2u + 2\sqrt{a^2 - u^2}\, \sin^{-1}\frac{u}{a}$

457. $\int \tan^{-1}\frac{u}{a}\, du = u \tan^{-1}\frac{u}{a} - \frac{a}{2} \ln(u^2 + a^2)$

458. $\int u \tan^{-1}\frac{u}{a}\, du = \frac{1}{2}(u^2 + a^2) \tan^{-1}\frac{u}{a} - \frac{au}{2}$

459. $\int u^2 \tan^{-1}\frac{u}{a}\, du = \frac{u^3}{3} \tan^{-1}\frac{u}{a} - \frac{au^2}{6} + \frac{a^3}{6} \ln(u^2 + a^2)$

460. $\int \frac{\tan^{-1}(u/a)}{u} du = \frac{u}{a} - \frac{(u/a)^3}{3^2} + \frac{(u/a)^5}{5^2} - \frac{(u/a)^7}{7^2} + \cdots$

461. $\int \frac{\tan^{-1}(u/a)}{u^2} du = -\frac{1}{u} \tan^{-1}\frac{u}{a} - \frac{1}{2a} \ln\left(\frac{u^2 + a^2}{u^2}\right)$

462. $\int \cot^{-1}\frac{u}{a}\, du = u \cot^{-1}\frac{u}{a} - \frac{a}{2} \ln(u^2 + a^2)$

463. $\int u \cot^{-1}\frac{u}{a}\, du = \frac{1}{2}(u^2 + a^2) \cot^{-1}\frac{u}{a} - \frac{au}{2}$

464. $\int u^2 \cot^{-1}\frac{u}{a}\, du = \frac{u^3}{3} \cot^{-1}\frac{u}{a} - \frac{au^2}{6} + \frac{a^3}{6} \ln(u^2 + a^2)$

465. $\int \frac{\cot^{-1}(u/a)}{u} du = \frac{\pi}{2} \ln|u| - \int \frac{\tan^{-1}(u/a)}{u} du$

466. $\int \frac{\cot^{-1}(u/a)}{u^2} du = -\frac{\cot^{-1}(u/a)}{u} + \frac{1}{2a} \ln\left(\frac{u^2 + a^2}{u^2}\right)$

467. $\int \sec^{-1}\frac{u}{a}\, du = \begin{cases} u \sec^{-1}\frac{u}{a} - a \ln\left|u + \sqrt{u^2 - a^2}\right|, & 0 < \sec^{-1}\frac{u}{a} < \frac{\pi}{2} \\ u \sec^{-1}\frac{u}{a} + a \ln\left|u + \sqrt{u^2 - a^2}\right|, & \frac{\pi}{2} < \sec^{-1}\frac{u}{a} < \pi \end{cases}$

468. $\displaystyle\int u \sec^{-1}\frac{u}{a}\,du = \begin{cases} \dfrac{u^2}{2}\sec^{-1}\dfrac{u}{a} - \dfrac{a\sqrt{u^2-a^2}}{2}, & 0 < \sec^{-1}\dfrac{u}{a} < \dfrac{\pi}{2} \\[3ex] \dfrac{u^2}{2}\sec^{-1}\dfrac{u}{a} + \dfrac{a\sqrt{u^2-a^2}}{2}, & \dfrac{\pi}{2} < \sec^{-1}\dfrac{u}{a} < \pi \end{cases}$

469. $\displaystyle\int u^2 \sec^{-1}\frac{u}{a}\,du = \begin{cases} \dfrac{u^3}{3}\sec^{-1}\dfrac{u}{a} - \dfrac{au\sqrt{u^2-a^2}}{6} - \dfrac{a^3}{6}\ln\left|u + \sqrt{u^2-a^2}\right|, & 0 < \sec^{-1}\dfrac{u}{a} < \dfrac{\pi}{2} \\[3ex] \dfrac{u^3}{3}\sec^{-1}\dfrac{u}{a} + \dfrac{au\sqrt{u^2-a^2}}{6} + \dfrac{a^3}{6}\ln\left|u + \sqrt{u^2-a^2}\right|, & \dfrac{\pi}{2} < \sec^{-1}\dfrac{u}{a} < \pi \end{cases}$

470. $\displaystyle\int \frac{\sec^{-1}(u/a)}{u}\,du = \frac{\pi}{a}\ln|u| + \frac{a}{u} + \frac{(u/a)^3}{2\cdot 3\cdot 3} + \frac{1\cdot 3(u/a)^5}{2\cdot 4\cdot 5\cdot 5} + \frac{1\cdot 3\cdot 5(u/a)^7}{2\cdot 4\cdot 6\cdot 7\cdot 7} + \cdots$

471. $\displaystyle\int \frac{\sec^{-1}(u/a)}{u}\,du = \begin{cases} -\dfrac{\sec^{-1}(u/a)}{u} + \dfrac{\sqrt{u^2-a^2}}{au}, & 0 < \sec^{-1}\dfrac{u}{a} < \dfrac{\pi}{2} \\[3ex] -\dfrac{\sec^{-1}(u/a)}{u} - \dfrac{\sqrt{u^2-a^2}}{au}, & \dfrac{\pi}{2} < \sec^{-1}\dfrac{u}{a} < \pi \end{cases}$

472. $\displaystyle\int \csc^{-1}\frac{u}{a}\,du = \begin{cases} u\csc^{-1}\dfrac{u}{a} + a\ln\left|u + \sqrt{u^2-a^2}\right|, & 0 < \csc^{-1}\dfrac{u}{a} < \dfrac{\pi}{2} \\[3ex] u\csc^{-1}\dfrac{u}{a} + a\ln\left|u + \sqrt{u^2-a^2}\right|, & \dfrac{\pi}{2} < \csc^{-1}\dfrac{u}{a} < 0 \end{cases}$

473. $\displaystyle\int u\csc^{-1}\frac{u}{a}\,du = \begin{cases} \dfrac{u^2}{2}\csc^{-1}\dfrac{u}{a} + \dfrac{a\sqrt{u^2-a^2}}{2}, & 0 < \csc^{-1}\dfrac{u}{a} < \dfrac{\pi}{2} \\[3ex] \dfrac{u^2}{2}\csc^{-1}\dfrac{u}{a} - \dfrac{a\sqrt{u^2-a^2}}{2}, & \dfrac{\pi}{2} < \csc^{-1}\dfrac{u}{a} < 0 \end{cases}$

474. $\displaystyle\int u^2\csc^{-1}\frac{u}{a}\,du = \begin{cases} \dfrac{u^3}{3}\csc^{-1}\dfrac{u}{a} - \dfrac{au\sqrt{u^2-a^2}}{6} - \dfrac{a^3}{6}\ln\left|u + \sqrt{u^2-a^2}\right|, & 0 < \csc^{-1}\dfrac{u}{a} < \dfrac{\pi}{2} \\[3ex] \dfrac{u^3}{3}\csc^{-1}\dfrac{u}{a} + \dfrac{au\sqrt{u^2-a^2}}{6} + \dfrac{a^3}{6}\ln\left|u + \sqrt{u^2-a^2}\right|, & \dfrac{\pi}{2} < \csc^{-1}\dfrac{u}{a} < 0 \end{cases}$

475. $\displaystyle\int \frac{\csc^{-1}(u/a)}{u}\,du = -\left(\frac{a}{u} + \frac{(a/u)^3}{2\cdot 3\cdot 3} + \frac{1\cdot 3(a/u)^5}{2\cdot 4\cdot 5\cdot 5} + \frac{1\cdot 3\cdot 5(a/u)^7}{2\cdot 4\cdot 6\cdot 7\cdot 7} + \cdots\right)$

476. $\displaystyle\int \frac{\csc^{-1}\frac{u}{a}}{u^2}\,du = \begin{cases} -\dfrac{\csc^{-1}\frac{u}{a}}{u} - \dfrac{\sqrt{u^2-a^2}}{au}, & 0 < \csc^{-1}\frac{u}{a} < \dfrac{\pi}{2} \\[3ex] -\dfrac{\csc^{-1}\frac{u}{a}}{u} + \dfrac{\sqrt{u^2-a^2}}{au}, & -\dfrac{\pi}{2} < \csc^{-1}\frac{u}{a} < 0 \end{cases}$

477. $\displaystyle\int u^n\sin^{-1}\frac{u}{a}\,du = \frac{u^{n+1}}{n+1}\sin^{-1}\frac{u}{a} - \frac{1}{n+1}\int \frac{u^{n+1}}{\sqrt{a^2-u^2}}\,du$

478. $\displaystyle\int u^n\cos^{-1}\frac{u}{a}\,du = \frac{u^{n+1}}{n+1}\cos^{-1}\frac{u}{a} + \frac{1}{n+1}\int \frac{u^{n+1}}{\sqrt{a^2-u^2}}\,du$

479. $\displaystyle\int u^n\tan^{-1}\frac{u}{a}\,du = \frac{u^{n+1}}{n+1}\tan^{-1}\frac{u}{a} - \frac{a}{n+1}\int \frac{u^{n+1}}{\sqrt{u^2+a^2}}\,du$

480. $\displaystyle\int u^n\cot^{-1}\frac{u}{a}\,du = \frac{u^{n+1}}{n+1}\cot^{-1}\frac{u}{a} - \frac{a}{n+1}\int \frac{u^{n+1}}{\sqrt{u^2+a^2}}\,du$

481. $\displaystyle\int u^n\sec^{-1}\frac{u}{a}\,du = \begin{cases} \dfrac{u^{n+1}\sec^{-1}(u/a)}{n+1} - \dfrac{a}{n+1}\displaystyle\int \dfrac{u^n\,du}{\sqrt{u^2-a^2}}, & 0 < \sec^{-1}\frac{u}{a} < \dfrac{\pi}{2} \\[3ex] \dfrac{u^{n+1}\sec^{-1}(u/a)}{n+1} + \dfrac{a}{n+1}\displaystyle\int \dfrac{u^n\,du}{\sqrt{u^2-a^2}}, & \dfrac{\pi}{2} < \sec^{-1}\frac{u}{a} < \pi \end{cases}$

482. $\displaystyle\int u^n\csc^{-1}\frac{u}{a}\,du = \begin{cases} \dfrac{u^{n+1}\csc^{-1}(u/a)}{n+1} + \dfrac{a}{n+1}\displaystyle\int \dfrac{u^n\,du}{\sqrt{u^2-a^2}} & 0 < \csc^{-1}\frac{u}{a} < \dfrac{\pi}{2} \\[3ex] \dfrac{u^{n+1}\csc^{-1}(u/a)}{n+1} - \dfrac{a}{n+1}\displaystyle\int \dfrac{u^n\,du}{\sqrt{u^2-a^2}} & -\dfrac{\pi}{2} < \csc^{-1}\frac{u}{a} < 0 \end{cases}$

INTEGRALS INVOLVING e^{au}

483. $\displaystyle\int e^{au}\,du = \frac{e^{au}}{a}$

484. $\displaystyle\int ue^{au}\,du = \frac{e^{au}}{a}\left(u - \frac{1}{a}\right)$

485. $\displaystyle\int u^2 e^{au}\,du = \frac{e^{au}}{a}\left(u^2 - \frac{2u}{a} + \frac{2}{a^2}\right)$

486. $\displaystyle\int u^n e^{au}\,du = \frac{u^n e^{au}}{a} - \frac{n}{a}\int u^{n-1} e^{au}\,du$

$$= \frac{e^{au}}{a}\left(u^n - \frac{nu^{n-1}}{a} + \frac{n(n-1)u^{n-2}}{a^2} - \cdots + \frac{(-1)^n n!}{a^n}\right), \quad \text{if } n = \text{positive integer}$$

487. $\displaystyle\int \frac{e^{au}}{u}\,du = \ln|u| + \frac{au}{1\cdot 1!} + \frac{(au)^2}{2\cdot 2!} + \frac{(au)^3}{3\cdot 3!} + \cdots$

488. $\displaystyle\int \frac{e^{au}}{u^n}\,du = \frac{-e^{au}}{(n-1)u^{n-1}} + \frac{a}{n-1}\int \frac{e^{au}}{u^{n-1}}\,du$

489. $\displaystyle\int \frac{du}{p+qe^{au}} = \frac{u}{p} - \frac{1}{ap}\ln\left|p+qe^{au}\right|$

490. $\displaystyle\int \frac{du}{(p+qe^{au})^2} = \frac{u}{p^2} + \frac{1}{ap(p+qe^{au})} - \frac{1}{ap^2}\ln\left|p+qe^{au}\right|$

491. $\displaystyle\int \frac{du}{pe^{au}+qe^{-au}} = \begin{cases} \dfrac{1}{a\sqrt{pq}}\tan^{-1}\left(\sqrt{\dfrac{p}{q}}\,e^{au}\right) \\[4mm] \dfrac{1}{2a\sqrt{-pq}}\ln\left|\dfrac{e^{au}-\sqrt{-q/p}}{e^{au}+\sqrt{-q/p}}\right| \end{cases}$

492. $\displaystyle\int e^{au}\sin bu\,du = \frac{e^{au}(a\sin bu - b\cos bu)}{a^2+b^2}$

493. $\displaystyle\int e^{au}\cos bu\,du = \frac{e^{au}(a\cos bu + b\sin bu)}{a^2+b^2}$

494. $\displaystyle\int u e^{au}\sin bu\,du = \frac{u e^{au}(a\sin bu - b\cos bu)}{a^2+b^2} - \frac{e^{au}\{(a^2-b^2)\sin bu - 2ab\cos bu\}}{(a^2+b^2)^2}$

495. $\displaystyle\int u e^{au}\cos bu\,du = \frac{u e^{au}(a\cos bu + b\sin bu)}{a^2+b^2} - \frac{e^{au}\{(a^2-b^2)\cos bu + 2ab\sin bu\}}{(a^2+b^2)^2}$

496. $\displaystyle\int e^{au}\ln|u|\,du = \frac{e^{au}\ln|u|}{a} - \frac{1}{a}\int \frac{e^{au}}{u}\,du$

497. $\displaystyle\int e^{au}\sin^n bu\,du = \frac{e^{au}\sin^{n-1} bu}{a^2+n^2b^2}(a\sin bu - nb\cos bu) + \frac{n(n-1)b^2}{a^2+n^2b^2}\int e^{au}\sin^{n-2} bu\,du$

498. $\displaystyle\int e^{au}\cos^n bu\,du = \frac{e^{au}\cos^{n-1} bu}{a^2+n^2b^2}(a\cos bu + nb\sin bu) + \frac{n(n-1)b^2}{a^2+n^2b^2}\int e^{au}\cos^{n-2} bu\,du$

INTEGRALS INVOLVING $\ln|u|$

499. $\displaystyle\int \ln|u|\,du = u\ln|u| - u$

500. $\displaystyle\int (\ln|u|)^2\,du = u(\ln|u|)^2 - 2u\ln|u| + 2u$

501. $\displaystyle\int (\ln|u|)^n\,du = u(\ln|u|)^n - n\int (\ln|u|)^{n-1}\,du$

502. $\displaystyle\int u\ln|u|\,du = \frac{u^2}{2}\left(\ln|u| - \frac{1}{2}\right)$

503. $\displaystyle\int u^m \ln|u|\,du = \frac{u^{m+1}}{m+1}\left(\ln|u| - \frac{1}{m+1}\right)\quad m \neq -1$

504. $\displaystyle\int \frac{\ln|u|}{u}\,du = \frac{1}{2}\ln^2|u|$

505. $\displaystyle\int \frac{\ln|u|}{u^2}\,du = -\frac{\ln|u|}{u} - \frac{1}{u}$

506. $\displaystyle\int \frac{(\ln|u|)^n\,du}{u} = \frac{(\ln|u|)^{n+1}}{n+1}$

507. $\displaystyle\int \frac{du}{u\ln|u|} = \ln|\ln|u||$

508. $\displaystyle\int \frac{du}{\ln|u|} = \ln|\ln|u|| + \ln|u| + \frac{(\ln|u|)^2}{2\cdot 2!} + \frac{(\ln|u|)^3}{3\cdot 3!} + \cdots$

509. $\displaystyle\int \frac{u^m\,du}{\ln|u|} = \ln|\ln|u|| + (m+1)\ln|u| + \frac{(m+1)^2(\ln|u|)^2}{2\cdot 2!} + \frac{(m+1)^3(\ln|u|)^3}{3\cdot 3!} + \cdots$

510. $\displaystyle\int u^m(\ln|u|)^n\,du = \frac{u^{m+1}(\ln|u|)}{m+1} - \frac{n}{m+1}\int u^m(\ln|u|)^{n-1}\,du$

511. $\displaystyle\int \ln(u^2 + a^2)\,du = u\ln(u^2 + a^2) - 2u + 2a\tan^{-1}\frac{u}{a}$

512. $\displaystyle\int \ln|u^2 - a^2|\,du = u\ln|u^2 - a^2| - 2u + a\ln\left|\frac{u+a}{u-2}\right|$

513. $\displaystyle\int u^m \ln|u^2 \pm a^2|\,du = \frac{u^{m+1}\ln|u^2 \pm a^2|}{m+1} - \frac{2}{m+1}\int \frac{u^{m+2}}{u^2 \pm a^2}\,du$

INTEGRALS INVOLVING cosh *au*

514. $\displaystyle\int \cosh au\ du = \frac{\sinh au}{a}$

515. $\displaystyle\int u \cosh au\ du = \frac{u \sinh au}{a} - \frac{\cosh au}{a^2}$

516. $\displaystyle\int u^2 \cosh au\ du = -\frac{2u \cosh au}{a^2} + \left(\frac{u^2}{a} + \frac{2}{a^3}\right) \sinh au$

517. $\displaystyle\int \frac{\cosh au}{u}\ du = \ln|u| + \frac{(ax)^2}{2 \cdot 2!} + \frac{(au)^4}{4 \cdot 4!} + \frac{(au)^6}{6 \cdot 6!} + \cdots$

518. $\displaystyle\int \frac{\cosh au}{u^2}\ du = -\frac{\cosh au}{u} + a \int \frac{\sinh au}{u}\ du$

519. $\displaystyle\int \frac{du}{\cosh au} = \frac{2}{a} \tan^{-1} e^{au}$

520. $\displaystyle\int \frac{u\ du}{\cosh au} = \frac{1}{a^2}\left\{\frac{(au)^2}{2} - \frac{(au)^4}{8} + \frac{5(au)^6}{144} + \cdots + \frac{(-1)^n E_n (au)^{2n+2}}{(2n+2)(2n)!} + \cdots\right\}$

521. $\displaystyle\int \cosh^2 au\ du = \frac{u}{2} + \frac{\sinh au \cosh au}{2a}$

522. $\displaystyle\int u \cosh^2 au\ du = \frac{u^2}{4} + \frac{u \sinh 2au}{4a} - \frac{\cosh 2au}{8a^2}$

523. $\displaystyle\int \frac{du}{\cosh^2 au} = \frac{\tanh au}{a}$

524. $\displaystyle\int \cosh au \cosh pu\ du = \frac{\sinh (a-p)u}{2(a-p)} + \frac{\sinh (a+p)u}{2(a+p)}$

525. $\displaystyle\int \cosh au \sin pu\ du = \frac{a \sinh au \sin pu - p \cosh au \cos pu}{a^2 + p^2}$

526. $\displaystyle\int \cosh au \cos pu\ du = \frac{a \sinh au \cos pu + p \cosh au \sin pu}{a^2 + p^2}$

527. $\displaystyle\int \frac{du}{\cosh au + 1} = \frac{1}{a} \tanh \frac{au}{2}$

528. $\displaystyle\int \frac{du}{\cosh au - 1} = -\frac{1}{a} \coth \frac{au}{2}$

529. $\int \dfrac{u \, du}{\cosh \, au + 1} = \dfrac{u}{a} \tanh \dfrac{au}{2} - \dfrac{2}{a^2} \ln \cosh \dfrac{au}{2}$

530. $\int \dfrac{u \, du}{\cosh \, au - 1} = -\dfrac{u}{a} \coth \dfrac{au}{2} + \dfrac{2}{a^2} \ln \left| \sinh \dfrac{au}{2} \right|$

531. $\int \dfrac{du}{(\cosh \, au + 1)^2} = -\dfrac{1}{2a} \tanh \dfrac{au}{2} - \dfrac{1}{6a} \tanh^3 \dfrac{au}{2}$

532. $\int \dfrac{du}{(\cosh \, au - 1)^2} = \dfrac{1}{2a} \coth \dfrac{au}{2} - \dfrac{1}{6a} \coth^3 \dfrac{au}{2}$

533. $\int \dfrac{du}{p + q \cosh \, au} = \begin{cases} \dfrac{2}{a\sqrt{q^2 - p^2}} \tan^{-1} \dfrac{qe^{au} + p}{\sqrt{q^2 - p^2}} \\[3ex] \dfrac{1}{a\sqrt{p^2 - q^2}} \ln \left| \dfrac{qe^{au} + p - \sqrt{p^2 - q^2}}{qe^{au} + p + \sqrt{p^2 - q^2}} \right| \end{cases}$

534. $\int \dfrac{du}{(p + q \cosh \, au)^2} = \dfrac{q \sinh \, au}{a(q^2 - p^2)(p + q \cosh \, au)} - \dfrac{p}{q^2 - p^2} \int \dfrac{du}{p + q \cosh \, au}$

535. $\int \dfrac{du}{p^2 - q^2 \cosh^2 \, au} = \begin{cases} \dfrac{1}{2ap\sqrt{p^2 - q^2}} \ln \left| \dfrac{p \tanh \, au + \sqrt{p^2 - q^2}}{p \tanh \, au - \sqrt{p^2 - q^2}} \right| \\[3ex] \dfrac{1}{ap\sqrt{q^2 - p^2}} \tan^{-1} \dfrac{p \tanh \, au}{\sqrt{q^2 - p^2}} \end{cases}$

536. $\int \dfrac{du}{p^2 + q^2 \cosh^2 \, au} = \begin{cases} \dfrac{1}{2ap\sqrt{p^2 + q^2}} \ln \left| \dfrac{p \tanh \, au + \sqrt{p^2 + q^2}}{p \tan \, au + \sqrt{q^2 - p^2}} \right| \\[3ex] \dfrac{1}{ap\sqrt{p^2 + q^2}} \tan^{-1} \dfrac{p \tanh \, au}{\sqrt{p^2 + q^2}} \end{cases}$

537. $\int u^m \cosh \, au \, du = -\dfrac{u^m \sinh \, au}{a} - \dfrac{m}{a} \int u^{m-1} \sinh \, au \, du$

538. $\int \cosh^n \, au \, du = \dfrac{\cosh^{n-1} \, au \sinh \, au}{an} + \dfrac{n-1}{n} \int \cosh^{n-2} \, au \, du$

539. $\int \dfrac{\cosh \, au}{u^n} \, du = \dfrac{-\cosh \, au}{(n-1)u^{n-1}} + \dfrac{a}{n-1} \int \dfrac{\sinh \, au}{u^{n-1}} \, du$

540. $\displaystyle\int \frac{du}{\cosh^n au} = \frac{\sinh au}{a(n-1)\cosh^{n-1} au} + \frac{n-2}{n-1} \int \frac{du}{\cosh^{n-2} au}$

541. $\displaystyle\int \frac{u\,du}{\cosh^n au} = \frac{u \sinh au}{a(n-1)\cosh^{n-1} au} + \frac{1}{(n-1)(n-2)a^2 \cosh^{n-2} au} + \frac{n-2}{n-1} \int \frac{u\,du}{\cosh^{n-2} au},$

INTEGRALS INVOLVING $\sinh au$

542. $\displaystyle\int \sinh au\,du = \frac{\cosh au}{a}$

543. $\displaystyle\int u \sinh au\,du = \frac{u \cosh au}{a} - \frac{\sinh au}{a^2}$

544. $\displaystyle\int u^2 \sinh au\,du = \left(\frac{u^2}{a} + \frac{2}{a^3}\right)\cosh au - \frac{2u}{a^2}\sinh au$

545. $\displaystyle\int \frac{\sinh au}{u}\,du = au + \frac{(au)^3}{3 \cdot 3!} + \frac{(au)^5}{5 \cdot 5!} + \cdots$

546. $\displaystyle\int \frac{\sinh au}{u^2}\,du = -\frac{\sinh au}{u} + a \int \frac{\cosh au}{u}\,du$

547. $\displaystyle\int \frac{du}{\sinh au} = \frac{1}{a}\ln\left|\tanh \frac{au}{2}\right|$

548. $\displaystyle\int \frac{u\,du}{\sinh au} = \frac{1}{a^2}\left\{au - \frac{(au)^3}{18} + \frac{7(au)^5}{1800} - \cdots + \frac{2(-1)^n(2^{2n}-1)B_n(au)^{2n+1}}{(2n+1)!} + \cdots\right\}$

549. $\displaystyle\int \sinh^2 au\,du = \frac{\sinh au \cosh au}{2a} - \frac{u}{2}$

550. $\displaystyle\int u \sinh^2 au\,du = \frac{u \sinh 2au}{4a} - \frac{\cosh 2au}{8a^2} - \frac{u^2}{4}$

551. $\displaystyle\int \frac{du}{\sinh^2 au} = -\frac{\coth au}{a}$

552. $\displaystyle\int \sinh au \sinh pu\,du = \frac{\sinh (a+p)u}{2(a+p)} - \frac{\sinh (a-p)u}{2(a-p)}$

553. $\displaystyle\int \sinh au \sin pu\,du = \frac{a \cosh au \sin pu - p \sinh au \cos pu}{a^2 + p^2}$

554. $\displaystyle\int \sinh au \cos pu\,du = \frac{a \cosh au \cos pu + p \sinh au \sin pu}{a^2 + p^2}$

555. $\int \dfrac{du}{p + q \sinh au} = \dfrac{1}{a\sqrt{p^2 + q^2}} \ln \left| \dfrac{qe^{au} + p - \sqrt{p^2 + q^2}}{qe^{au} + p + \sqrt{p^2 + q^2}} \right|$

556. $\int \dfrac{du}{(p + q \sinh au)^2} = \dfrac{- q \cosh au}{a(p^2 + q^2)(p + q \sinh au)} + \dfrac{p}{p^2 + q^2} \int \dfrac{du}{p + q \sinh au}$

557. $\int \dfrac{du}{p^2 + q^2 \sinh^2 au} = \begin{cases} \dfrac{1}{ap\sqrt{q^2 - p^2}} \tan^{-1} \dfrac{\sqrt{q^2 - p^2}\ \tanh au}{p} \\[4mm] \dfrac{1}{2ap\sqrt{p^2 - q^2}} \ln \left| \dfrac{p + \sqrt{p^2 - q^2}\ \tanh au}{p - \sqrt{p^2 - q^2}\ \tanh au} \right| \end{cases}$

558. $\int \dfrac{du}{p^2 - q^2 \sinh^2 au} = \dfrac{1}{2ap\sqrt{p^2 + q^2}} \ln \left| \dfrac{p + \sqrt{p^2 + q^2}\ \tanh au}{p - \sqrt{p^2 + q^2}\ \tanh au} \right|$

559. $\int u^m \sinh au\ du = \dfrac{u^m \cosh au}{a} - \dfrac{m}{a} \int u^{m-1} \cosh au\ du$

560. $\int \sinh^n au\ du = \dfrac{\sinh^{n-1} au \cosh au}{an} - \dfrac{n-1}{n} \int \sinh^{n-2} au\ du$

561. $\int \dfrac{\sinh au}{u^n} du = - \dfrac{- \sinh au}{(n-1)u^{n-1}} + \dfrac{a}{n-1} \int \dfrac{\cosh au}{u^{n-1}} du$

562. $\int \dfrac{du}{\sinh^n au} = \dfrac{- \cosh au}{a(n-1)\sinh^{n-1} au} - \dfrac{n-2}{n-1} \int \dfrac{du}{\sinh^{n-2} au}$

563. $\int \dfrac{u\ du}{\sinh^n au} = \dfrac{- u \cosh au}{a(n-1)\sinh^{n-1} au} - \dfrac{1}{a^2(n-1)(n-2)\sinh^{n-2} au} - \dfrac{n-2}{n-1} \int \dfrac{u\ du}{\sinh^{n-2} au}.$

INTEGRALS INVOLVING sinh *au* AND cosh *au*

564. $\int \sinh au \cosh au\ du = \dfrac{\sinh^2 au}{2a}$

565. $\int \sinh pu \cosh qu\ du = \dfrac{\cosh (p + q)u}{2(p + q)} + \dfrac{\cosh (p - q)u}{2(p - q)}$

566. $\int \sinh^n au \cosh au\ du = \dfrac{\sinh^{n+1} au}{(n + 1)a}$

567. $\displaystyle\int \cosh^n au \, \sinh \, au \, du = \frac{\cosh^{n+1} au}{(n+1)a}$

568. $\displaystyle\int \sinh^2 au \, \cosh^2 au \, du = \frac{\sinh 4au}{32a} - \frac{u}{8}$

569. $\displaystyle\int \frac{du}{\sinh au \, \cosh au} = \frac{1}{a} \ln|\tanh \, au|$

570. $\displaystyle\int \frac{du}{\sinh^2 au \, \cosh au} = -\frac{1}{a} \tan^{-1} \sinh \, au - \frac{\operatorname{csch} au}{a}$

571. $\displaystyle\int \frac{du}{\sinh au \, \cosh^2 au} = \frac{\operatorname{sech}}{a} + \frac{1}{a} \ln\left|\tanh \frac{au}{2}\right|$

572. $\displaystyle\int \frac{du}{\sinh^2 au \, \cosh^2 au} = -\frac{2 \coth 2au}{a}$

573. $\displaystyle\int \frac{\sinh^2 au}{\cosh au} \, du = \frac{\sinh au}{a} - \frac{1}{a} \tan^{-1} \sinh \, au$

574. $\displaystyle\int \frac{\cosh^2 au}{\sinh au} \, du = \frac{\cosh au}{a} + \frac{1}{a} \ln\left|\tanh \frac{au}{2}\right|$

575. $\displaystyle\int \frac{du}{\cosh au \, (1 + \sinh au)} = \frac{1}{2a} \ln\left|\frac{1 + \sinh au}{\cosh au}\right| + \frac{1}{a} \tan^{-1} e^{au}$

576. $\displaystyle\int \frac{du}{\sinh au \, (\cosh au + 1)} = \frac{1}{2a} \ln\left|\tanh \frac{au}{2}\right| + \frac{1}{2a(\cosh au + 1)}$

577. $\displaystyle\int \frac{du}{\cosh au \, (1 + \sinh au)} = \frac{1}{2a} \ln\left|\tanh \frac{au}{2}\right| - \frac{1}{2a(\cosh au - 1)}$

INTEGRALS INVOLVING tanh *au*

578. $\displaystyle\int \tanh \, au \, du = \frac{1}{a} \ln \cosh \, au$

579. $\displaystyle\int \tanh^2 au \, du = u - \frac{\tanh au}{a}$

580. $\displaystyle\int \tanh^3 au \, du = \frac{1}{a} \ln \cosh \, au - \frac{\tanh^2 au}{2a}$

581. $\displaystyle\int \tanh^n au \, \operatorname{sech}^2 au \, du = \frac{\tanh^{n+1} au}{(n+1)a}$

582. $\int \dfrac{\text{sech}^2 \, au}{\tanh \, au} \, du = -\dfrac{1}{a} \ln|\tanh \, au|$

583. $\int \dfrac{du}{\tanh \, au} = -\dfrac{1}{a} \ln|\sinh \, au|$

584. $\int u \tanh \, au \, du = \dfrac{1}{a^2}\left\{ \dfrac{(au)^3}{3} - \dfrac{(au)^5}{15} + \dfrac{2(au)^7}{105} - \cdots \dfrac{(-1)^{n-1} 2^{2n}(2^{2n}-1)B_n (au)^{2n+1}}{(2n+1)!} + \cdots \right\}$

585. $\int u \tanh^2 \, au \, du = \dfrac{u^2}{2} - \dfrac{u \tanh \, au}{a} + \dfrac{1}{a^2} \ln \cosh \, au$

586. $\int \dfrac{\tanh \, au}{u} \, du = au - \dfrac{(au)^3}{9} + \dfrac{2(au)^5}{75} - \cdots \dfrac{(-1)^{n-1} 2^{2n}(2^{2n}-1)B_n (au)^{2n+1}}{(2n-1)!(2n)!} + \cdots$

587. $\int \dfrac{du}{p + q \tanh \, au} = \dfrac{pu}{p^2 - q^2} - \dfrac{q}{a(p^2 - q^2)} \ln|p \sinh \, au + q \cosh \, au|$

588. $\int \tanh^n \, au \, du = \dfrac{-\tanh^{n-1} \, au}{a(n-1)} + \int \tanh^{n-2} \, au \, du$

INTEGRALS INVOLVING coth *au*

589. $\int \coth \, au \, du = \dfrac{1}{a} \ln|\sinh \, au|$

590. $\int \coth^2 \, au \, du = u - \dfrac{\coth \, au}{a}$

591. $\int \coth^3 \, au \, du = \dfrac{1}{a} \ln|\sinh \, au| - \dfrac{\coth^2 \, au}{2a}$

592. $\int \coth^n \, au \, \text{csch}^2 \, au \, du = \dfrac{\coth^{n+1} au}{(n+1)a}$

593. $\int \dfrac{\text{csch}^2 \, au}{\coth \, au} \, du = -\dfrac{1}{a} \ln|\coth \, au|$

594. $\int \dfrac{du}{\coth \, au} = -\dfrac{1}{a} \ln(\cosh \, au)$

595. $\int u \coth \, au \, du = \dfrac{1}{a^2}\left\{ au + \dfrac{(au)^3}{9} - \dfrac{(au)^5}{225} + \cdots \dfrac{(-1)^{n-1} 2^{2n} B_n (au)^{2n+1}}{(2n+1)!} + \cdots \right\}$

596. $\int u \coth^2 \, au \, du = \dfrac{u^2}{2} - \dfrac{u \coth \, au}{a} + \dfrac{1}{a^2} \ln|\sinh \, au|$

597. $\int \dfrac{\coth au}{u}\, du = -\dfrac{1}{au} + \dfrac{au}{3} - \dfrac{(au)^3}{135} + \cdots \dfrac{(-1)^n 2^{2n} B_n (au)^{2n-1}}{(2n-1)(2n)!} + \cdots$

598. $\int \dfrac{du}{p + q \coth au} = \dfrac{pu}{p^2 - q^2} - \dfrac{q}{a(p^2 - q^2)} \ln|\, p \sinh au + q \cosh au\,|$

599. $\int \coth^n au\, du = -\dfrac{\coth^{n-1} au}{a(n-1)} + \int \coth^{n-2} au\, du$

INTEGRALS INVOLVING sech *au*

600. $\int \operatorname{sech} au\, du = \dfrac{2}{a} \ln|\tan^{-1} e^{au}|$

601. $\int \operatorname{sech}^2 au\, du = \dfrac{\tanh au}{a}$

602. $\int \operatorname{sech}^3 au\, du = \dfrac{\operatorname{sech} au \tanh au}{2a} + \dfrac{1}{2a} \tan^{-1} \sinh au$

603. $\int \operatorname{sech}^n au \tanh au\, du = -\dfrac{\operatorname{sech}^n au}{na}$

604. $\int \dfrac{du}{\operatorname{sech} au} = \dfrac{\sinh au}{a}$

605. $\int u \operatorname{sech} au\, du = \dfrac{1}{a^2} \left\{ \dfrac{(au)^2}{2} - \dfrac{(au)^4}{8} + \dfrac{5(au)^6}{144} + \cdots \dfrac{(-1)^n E_n (au)^{2n+2}}{(2n+2)(2n)!} + \cdots \right\}$

606. $\int u \operatorname{sech}^2 au\, du = -\dfrac{u \tanh au}{a} - \dfrac{1}{a^2} \ln \cosh au$

607. $\int \dfrac{\operatorname{sech} au}{u}\, du = \ln|\,u\,| - \dfrac{(au)^2}{4} + \dfrac{5(au)^4}{96} - \dfrac{61(au)^6}{4320} + \cdots \dfrac{(-1)^n E_n (au)^{2n}}{2n(2n)!} + \cdots$

608. $\int \dfrac{du}{q + p \operatorname{sech} au} = \dfrac{u}{q} - \dfrac{p}{q} \int \dfrac{du}{p + q \cosh au}$

609. $\int \operatorname{sech}^n au\, du = \dfrac{\operatorname{sech}^{n-2} au \tanh au}{a(n-1)} + \dfrac{n-2}{n-1} \int \operatorname{sech}^{n-2} au\, du$

INTEGRALS INVOLVING csch *au*

610. $\int \operatorname{csch} au\, du = \dfrac{1}{a} \ln\left|\tanh \dfrac{au}{2}\right|$

611. $\displaystyle\int \operatorname{csch}^2 au \; du = -\frac{\coth au}{a}$

612. $\displaystyle\int \operatorname{csch}^3 au \; du = -\frac{\operatorname{csch} au \coth au}{2a} - \frac{1}{2a} \ln\left|\tanh \frac{au}{2}\right|$

613. $\displaystyle\int \operatorname{csch}^n au \coth au \; du = -\frac{\operatorname{csch}^n au}{na}$

614. $\displaystyle\int \frac{du}{\operatorname{csch} au} = \frac{1}{a} \cosh au$

615. $\displaystyle\int u \operatorname{csch} au \; du = \frac{1}{a^2}\left\{ au - \frac{(au)^3}{18} + \frac{7(au)^5}{1800} + \cdots + \frac{2(-1)^n(2^{2n-1}-1)B_n(au)^{2n+1}}{(2n+1)!} + \cdots \right\}$

616. $\displaystyle\int u \operatorname{csch}^2 au \; du = -\frac{u \operatorname{csch} au}{a} + \frac{1}{a^2}\ln|\sinh au|$

617. $\displaystyle\int \frac{\operatorname{csch} au}{u} du = -\frac{1}{au} - \frac{au}{6} + \frac{7(au)^3}{1080} + \cdots \frac{(-1)^n 2(2^{2n}-1)B_n(au)^{2n-1}}{(2n-1)(2n)!} + \cdots$

618. $\displaystyle\int \frac{du}{q + p \operatorname{csch} au} = \frac{u}{q} - \frac{p}{q}\int \frac{du}{p + q \sinh au}$

619. $\displaystyle\int \operatorname{csch}^n au \; du = \frac{-\operatorname{csch}^{n-2} au \coth au}{a(n-1)} - \frac{n-2}{n-1}\int \operatorname{csch}^{n-2} au \; du$

INTEGRALS INVOLVING INVERSE HYPERBOLIC FUNCTIONS

620. $\displaystyle\int \sinh^{-1}\frac{u}{a} \; du = u \sinh^{-1}\frac{u}{a} - \sqrt{u^2 + a^2}$

621. $\displaystyle\int u \sinh^{-1}\frac{u}{a} \; du = \left(\frac{u^2}{2} + \frac{a^2}{4}\right)\sinh^{-1}\frac{u}{a} - \frac{u\sqrt{u^2 + a^2}}{4}$

622. $\displaystyle\int u^2 \sinh^{-1}\frac{u}{a} \; du = \frac{u^3}{3}\sinh^{-1}\frac{u}{a} + \frac{(2a^2 - u^2)\sqrt{u^2 + a^2}}{9}$

623. $\displaystyle\int \frac{\sinh^{-1}(u/a)}{u} du \begin{cases} \dfrac{u}{a} - \dfrac{(u/a)^3}{2\cdot3\cdot3} + \dfrac{1\cdot3(u/a)^5}{2\cdot4\cdot5\cdot5} - \dfrac{1\cdot3\cdot5(u/a)^7}{2\cdot4\cdot6\cdot7\cdot7} + \cdots, & |u| < a \\[2mm] \dfrac{\ln^2|2u/a|}{2} - \dfrac{(a/u)^2 3}{2\cdot2\cdot2} - \dfrac{1\cdot3(a/u)^4}{2\cdot4\cdot4\cdot4} - \dfrac{1\cdot3\cdot5(a/u)^6}{2\cdot4\cdot6\cdot6\cdot6} + \cdots, & u > a \\[2mm] -\dfrac{\ln^2|-2u/a|}{2} + \dfrac{(a/u)^2}{2\cdot2\cdot2} - \dfrac{1\cdot3(a/u)^4}{2\cdot4\cdot4\cdot4} + \dfrac{1\cdot3\cdot5(a/u)^6}{2\cdot4\cdot6\cdot6\cdot6} - \cdots, & u > -a \end{cases}$

624. $\displaystyle\int \frac{\sin^{-1}(u/a)}{u^2}\, du = -\frac{\sinh^{-1}(u/a)}{u} - \frac{1}{a}\ln\left|\frac{a + \sqrt{u^2 + a^2}}{u}\right|$

625. $\displaystyle\int \cosh^{-1}\frac{u}{a}\, du = \begin{cases} u\cosh^{-1}(u/a) - \sqrt{u^2 - a^2}, & \cosh^{-1}(u/a) > 0 \\ u\cosh^{-1}(u/a) + \sqrt{u^2 - a^2}, & \cosh^{-1}(u/a) < 0 \end{cases}$

626. $\displaystyle\int u\cosh^{-1}\frac{u}{a}\, du = \begin{cases} \frac{1}{4}(2u^2 - a^2)\cosh^{-1}(u/a) - \frac{1}{4}u\sqrt{u^2 - a^2}, & \cosh^{-1}(u/a) > 0 \\ \frac{1}{4}(2u^2 - a^2)\cosh^{-1}(u/a) + \frac{1}{4}u\sqrt{u^2 - a^2}, & \cosh^{-1}(u/a) < 0 \end{cases}$

627. $\displaystyle\int u^2\cosh^{-1}\frac{u}{a}\, du = \begin{cases} \frac{1}{3}u^3\cosh^{-1}(u/a) - \frac{1}{9}(u^2 + 2a^2)\sqrt{u^2 - a^2}, & \cosh^{-1}(u/a) > 0 \\ \frac{1}{3}u^3\cosh^{-1}(u/a) + \frac{1}{9}(u^2 + 2a^2)\sqrt{u^2 - a^2}, & \cosh^{-1}(u/a) < 0 \end{cases}$

628. $\displaystyle\int \frac{\cosh^{-1}(u/a)}{u}\, du = \pm\left[\frac{1}{2}\ln^2|2u/a| + \frac{(a/u)^2}{2\cdot 2\cdot 2} + \frac{1\cdot 3(a/u)^4}{2\cdot 4\cdot 4\cdot 4} + \frac{1\cdot 3\cdot 5(a/u)^6}{2\cdot 4\cdot 6\cdot 6\cdot 6} + \cdots\right]$

$+$ if $\cosh^{-1}(u/a) > 0$, $-$ if $\cosh^{-1}(u/a) < 0$

629. $\displaystyle\int \frac{\cosh^{-1}(u/a)}{u^2}\, du = -\frac{\cosh^{-1}(u/a)}{u} \mp \frac{1}{a}\ln\left|\frac{a + \sqrt{u^2 + a^2}}{u}\right|$

$-$ if $\cosh^{-1}(u/a) > 0$, $+$ if $\cosh^{-1}(u/a) < 0$

630. $\displaystyle\int \tanh^{-1}\frac{u}{a}\, du = u\tanh^{-1}\frac{u}{a} + \frac{a}{2}\ln|a^2 - u^2|$

631. $\displaystyle\int u\tanh^{-1}\frac{u}{a}\, du = \frac{au}{2} + \frac{1}{2}(u^2 - a^2)\tanh^{-1}\frac{u}{a}$

632. $\displaystyle\int u^2\tanh^{-1}\frac{u}{a}\, du = \frac{au^2}{6} + \frac{u^3}{3}\tanh^{-1}\frac{u}{a} + \frac{a^3}{6}\ln|a^2 - u^2|$

633. $\displaystyle\int \frac{\tanh^{-1}(u/a)}{u}\, du = \frac{u}{a} + \frac{(u/a)^3}{3^2} + \frac{(u/a)^5}{5^2} + \cdots$

634. $\displaystyle\int \frac{\tanh^{-1}(u/a)}{u^2}\, du = -\frac{\tanh^{-1}}{u} + \frac{1}{2a}\ln\left|\frac{u^2}{a^2 - u^2}\right|$

635. $\displaystyle\int \coth^{-1}\frac{u}{a}\, du = u\coth^{-1}u - \frac{a}{2}\ln|u^2 - a^2|$

636. $\displaystyle\int u\coth^{-1}\frac{u}{a}\, du = \frac{au}{2} + \frac{1}{2}(u^2 - a^2)\coth^{-1}\frac{u}{a}$

637. $\displaystyle\int u^2\coth^{-1}\frac{u}{a}\, du = \frac{au^2}{6} + \frac{u^3}{3}\coth^{-1}\frac{u}{a} + \frac{a^3}{6}\ln|u^2 - a^2|$

638. $\displaystyle\int \frac{\coth^{-1}(u/a)}{u}\, du = -\left(\frac{a}{u} + \frac{(a/u)^3}{3^2} + \frac{(a/u)^5}{5^2} + \cdots\right)$

639. $\displaystyle\int \frac{\coth^{-1}(u/a)}{u^2}\,du = -\frac{\coth^{-1}(u/a)}{u} + \frac{1}{2a}\ln\left|\frac{u^2}{u^2 - a^2}\right|$

640. $\displaystyle\int \operatorname{sech}^{-1}\frac{u}{a}\,du = \begin{cases} u\operatorname{sech}^{-1}(u/a) + a\sin^{-1}(u/a), & \operatorname{sech}^{-1}(u/a) > 0 \\[2mm] u\operatorname{sech}^{-1}(u/a) - a\sin^{-1}(u/a), & \operatorname{sech}^{-1}(u/a) < 0 \end{cases}$

641. $\displaystyle\int u\operatorname{sech}^{-1}\frac{u}{a}\,du = \begin{cases} \frac{1}{2}u^2\operatorname{sech}^{-1}(u/a) - \frac{1}{2}a\sqrt{a^2 - u^2}, & \operatorname{sech}^{-1}(u/a) > 0 \\[2mm] \frac{1}{2}u^2\operatorname{sech}^{-1}(u/a) + \frac{1}{2}a\sqrt{a^2 - u^2}, & \operatorname{sech}^{-1}(u/a) < 0 \end{cases}$

642. $\displaystyle\int \frac{\operatorname{sech}^{-1}}{u}\,du = \begin{cases} -\frac{1}{2}\ln\left|\frac{a}{u}\right|\ln\left|\frac{4a}{u}\right| - \frac{(u/a)^2}{2\cdot 2\cdot 2} - \frac{1\cdot 3(u/a)^4}{2\cdot 4\cdot 4\cdot 4} - \cdots, & \operatorname{sech}^{-1}(u/a) > 0 \\[2mm] \frac{1}{2}\ln\left|\frac{a}{u}\right|\ln\left|\frac{4a}{u}\right| + \frac{(u/a)^2}{2\cdot 2\cdot 2} - \frac{1\cdot 3(u/a)^4}{2\cdot 4\cdot 4\cdot 4} + \cdots, & \operatorname{sech}^{-1}(u/a) < 0 \end{cases}$

643. $\displaystyle\int \operatorname{csch}^{-1}\frac{u}{a}\,du = u\operatorname{csch}^{-1}\frac{u}{a} \pm a\sinh^{-1}\frac{u}{a}$ ["+" if $u > 0$, "−" if $u < 0$]

644. $\displaystyle\int u\operatorname{csch}^{-1}\frac{u}{a}\,du = \frac{u^2}{2}\operatorname{csch}^{-1}\frac{u}{a} \pm \frac{a\sqrt{u^2 + a^2}}{2}$ ["+" if $u > 0$, "−" if $u < 0$]

645. $\displaystyle\int \frac{\operatorname{csch}^{-1}(u/a)}{u}\,du = \begin{cases} \frac{1}{2}\ln\left|\frac{a}{u}\right|\ln\left|\frac{4a}{u}\right| + \frac{1(u/a)^2}{2\cdot 2\cdot 2} - \frac{1\cdot 3(u/a)^4}{2\cdot 4\cdot 4\cdot 4} + \cdots, & 0 < u < a \\[2mm] \frac{1}{2}\ln\left|-\frac{u}{a}\right|\ln\left|-\frac{u}{4a}\right| - \frac{(u/a)^2}{2\cdot 2\cdot 2} + \frac{1\cdot 3(u/a)^4}{2\cdot 4\cdot 4\cdot 4} - \cdots, & -a < u < 0 \\[2mm] -\frac{a}{u} + \frac{(a/u)^3}{2\cdot 3\cdot 3} - \frac{1\cdot 3(a/u)^5}{2\cdot 4\cdot 5\cdot 5} + \cdots, & |u| > a \end{cases}$

645. $\displaystyle\int u^m\sinh^{-1}\frac{u}{a}\,du = \frac{u^{m+1}}{m+1}\sinh^{-1}\frac{u}{a} - \frac{1}{m+1}\int \frac{u^{m+1}}{\sqrt{u^2 + a^2}}\,du$

646. $\displaystyle\int u^m\cosh^{-1}\frac{u}{a}\,du = \begin{cases} \frac{u^{m+1}}{m+1}\cosh^{-1}\frac{u}{a} - \frac{1}{m+1}\int \frac{u^{m+1}}{\sqrt{u^2 - a^2}}\,du, & \cosh^{-1}(u/a) > 0 \\[2mm] \frac{u^{m+1}}{m+1}\cosh^{-1}\frac{u}{a} + \frac{1}{m+1}\int \frac{u^{m+1}}{\sqrt{u^2 - a^2}}\,du, & \cosh^{-1}(u/a) < 0 \end{cases}$

647. $\displaystyle\int u^m\tanh^{-1}\frac{u}{a}\,du = \frac{u^{m+1}}{m+1}\tanh^{-1}\frac{u}{a} - \frac{a}{m+1}\int \frac{u^{m+1}}{a^2 - u^2}\,du$

648. $\displaystyle\int u^m\coth^{-1}\frac{u}{a}\,du = \frac{u^{m+1}}{m+1}\coth^{-1}\frac{u}{a} - \frac{a}{m+1}\int \frac{u^{m+1}}{a^2 - u^2}\,du$

649. $\displaystyle\int u^m \text{sech}^{-1}\frac{u}{a}\,du = \begin{cases} \dfrac{u^{m+1}}{m+1}\,\text{sech}^{-1}\dfrac{u}{a} + \dfrac{a}{m+1}\displaystyle\int\dfrac{u^m\,du}{\sqrt{a^2-u^2}}, & \text{sech}^{-1}(u/a) > 0 \\[4mm] \dfrac{u^{m+1}}{m+1}\,\text{sech}^{-1}\dfrac{u}{a} - \dfrac{a}{m+1}\displaystyle\int\dfrac{u^m\,du}{\sqrt{a^2-u^2}}, & \text{sech}^{-1}(u/a) < 0 \end{cases}$

650. $\displaystyle\int u^m \text{csch}^{-1}\frac{u}{a}\,du = \frac{u^{m+1}}{m+1}\,\text{csch}^{-1}\frac{u}{a} \pm \frac{a}{m+1}\int\frac{u^m\,du}{\sqrt{u^2+a^2}},$ \quad ["+" if $u>0$, "$-$" if $u<0$]

10.5

Bernoulli and Euler Numbers; Gamma and Beta Functions

Bernoulli and Euler Numbers

The *Bernoulli numbers* B_1, B_2, B_3, $\cdots$ are defined by the series

$$\frac{x}{e^x - 1} = 1 - \frac{x}{2} + \frac{B_1 x^2}{2!} - \frac{B_2 x^4}{4!} + \frac{B_3 x^6}{6!} - \cdots \quad |x| < 2\pi$$

The *Euler numbers* E_1, E_2, E_3, $\cdots$ are defined by the series

$$\sec x = 1 + \frac{E_1 x^2}{2!} + \frac{E_2 x^4}{4!} + \frac{E_3 x^6}{6!} + \cdots \quad |x| < \frac{\pi}{2}$$

TABLE OF FIRST FEW BERNOULLI AND EULER NUMBERS

Bernoulli numbers	Euler numbers
$B_1 = 1/6$	$E_1 = 1$
$B_2 = 1/30$	$E_2 = 5$
$B_3 = 1/42$	$E_3 = 61$
$B_4 = 1/30$	$E_4 = 1{,}385$
$B_5 = 5/66$	$E_5 = 50{,}521$
$B_6 = 691/2730$	$E_6 = 2{,}702{,}765$
$B_7 = 7/6$	$E_7 = 199{,}360{,}981$
$B_8 = 3617/510$	$E_8 = 19{,}391{,}512{,}145$
$B_9 = 43{,}867/798$	$E_9 = 2{,}404{,}879{,}675{,}441$
$B_{10} = 174{,}611/330$	$E_{10} = 370{,}371{,}188{,}237{,}525$
$B_{11} = 854{,}513/138$	$E_{11} = 69{,}348{,}874{,}393{,}137{,}901$
$B_{12} = 236{,}364{,}091/2730$	$E_{12} = 15{,}514{,}534{,}163{,}557{,}086{,}905$

GAMMA FUNCTION

For $n > 0$, $\Gamma(n) = \displaystyle\int_0^\infty t^{n-1} e^{-t}\, dt; \quad \Gamma(n+1) = n\Gamma(n)$

For $n = 0, 1, 2, \cdots$ $\Gamma(n+1) = n!$

For $n < 0$ $\Gamma(n) = \dfrac{\Gamma(n+1)}{n}$

BETA FUNCTION

$$B(m, n) = \int_0^1 t^{m-1}(1 - t)^{n-1}\, dt \quad\quad m > 0,\, n > 0$$

$$B(m, n) = \frac{\Gamma(m)\Gamma(n)}{\Gamma(m + n)}$$

10.6

Definite Integral Formulas

DEFINITE INTEGRAL
AT A POINT
or EQUAL LIMITS

1. $\displaystyle\int_a^a f(x)\,dx = 0$

OPPOSITE OF A
DEFINITE INTEGRAL
or LIMIT REVERSAL

2. $\displaystyle\int_a^b f(x)\,dx = -\int_b^a f(x)\,dx$

SUM/
DIFFERENCE

3. $\displaystyle\int_a^b \{f(x) \pm g(x)\}\,dx = \int_a^b f(x)\,dx \pm \int_a^b g(x)\,dx$

CONSTANT

4. $\displaystyle\int_a^b cf(x)\,dx = c\int_a^b f(x)\,dx$ where c is any constant

INTERIOR
POINT
(subdivision rule)

5. $\displaystyle\int_a^b f(x)\,dx = \int_a^c f(x)\,dx + \int_c^b f(x)\,dx$

MEAN VALUE
THEOREM

6. $\displaystyle\int_a^b f(x)\,dx = (b-a)f(c)$ where c is between a and b

IMPROPER
INTEGRALS

7. $\displaystyle\int_a^\infty f(x)\,dx = \lim_{b\to\infty}\int_a^b f(x)\,dx$

8. $\displaystyle\int_{-\infty}^\infty f(x)\,dx = \lim_{a\to-\infty}\int_a^c f(x)\,dx + \lim_{b\to\infty}\int_c^b f(x)\,dx$

9. $\displaystyle\int_a^b f(x)\,dx = \lim_{\epsilon\to0}\int_a^{b-\epsilon} f(x)\,dx$ if b is a singular point

10. $\displaystyle\int_a^b f(x)\,dx = \lim_{\epsilon\to0}\int_{a+\epsilon}^b f(x)\,dx$ if a is a singular point

11. $\displaystyle\int_a^b f(x)\,g(x)\,dx = f(c)\int_a^b g(x)\,dx$ where c is between a and b

This is a generalization of formula 6 and is valid if $f(x)$ and $g(x)$ are continuous in $a \le x \le b$ and $g(x) \ge 0$.

APPROXIMATION FORMULAS FOR DEFINITE INTEGRALS

In the following the interval from $x = a$ to $x = b$ is subdivided into n equal parts by the points $a = x_0, x_1, x_2, \cdots, x_{n-1}, x_n = b$ and we let $y_0 = f(x_0)$, $y_1 = f(x_1)$, $y_2 = f(x_2)$, $\cdots$, $y_n = f(x_n)$, $h = (b - a)/n$.

Rectangular formula

12. $\displaystyle\int_a^b f(x)\, dx \approx h(y_0 + y_1 + y_2 + \cdots + y_{n-1})$

Trapezoidal formula

13. $\displaystyle\int_a^b f(x)\, dx \approx \frac{h}{2}(y_0 + 2y_1 + 2y_2 + \cdots + 2y_{n-1} + y_n)$

Simpson's formula (or parabolic formula) for n even

14. $\displaystyle\int_a^b f(x)\, dx \approx \frac{h}{3}(y_0 + 4y_1 + 2y_2 + 4y_3 + \cdots + 2y_{n-2} + 4y_{n-1} + y_n)$

DEFINITE INTEGRALS INVOLVING RATIONAL OR IRRATIONAL EXPRESSIONS

15. $\displaystyle\int_0^\infty \frac{dx}{x^2 + a^2} = \frac{\pi}{2a}$

16. $\displaystyle\int_0^\infty \frac{x^{p-1}\, dx}{1 + x} = \frac{\pi}{\sin p\pi}, \quad 0 < p < 1$

17. $\displaystyle\int_0^\infty \frac{x^m\, dx}{x^n + a^n} = \frac{\pi a^{m+1-n}}{n \sin \left[(m+1)\pi/n\right]}, \quad 0 < m + 1 < n$

18. $\displaystyle\int_0^\infty \frac{x^m\, dx}{1 + 2x \cos \beta + x^2} = \frac{\pi}{\sin m\pi} \frac{\sin m\beta}{\sin \beta}$

19. $\displaystyle\int_0^a \frac{dx}{\sqrt{a^2 - x^2}} = \frac{\pi}{2}$

20. $\displaystyle\int_0^a \sqrt{a^2 - x^2}\, dx = \frac{\pi a^2}{4}$

21. $\displaystyle\int_0^a x^m(a^n - x^n)^p \, dx = \frac{a^{m+1+np}\Gamma[(m+1)/n]\Gamma(p+1)}{n\Gamma[(m+1)/n + p + 1]}$

22. $\displaystyle\int_0^\infty \frac{x^m \, dx}{(x^n + a^n)^r} = \frac{(-1)^{r-1}\pi a^{m+1-nr}\Gamma[(m+1)/n]}{n\sin[(m+1)\pi/n](r-1)!\Gamma[(m+1)/n - r + 1]}, \quad 0 < m+1 < nr$

DEFINITE INTEGRALS INVOLVING TRIGONOMETRIC FUNCTIONS

All variables are considered positive unless otherwise indicated.

23. $\displaystyle\int_0^\pi \sin mx \sin nx \, dx = \begin{cases} 0, & m, n \text{ integers and } m \neq n \\ \pi/2, & m, n \text{ integers and } m = n \end{cases}$

24. $\displaystyle\int_0^\pi \cos mx \cos nx \, dx = \begin{cases} 0, & m, n \text{ integers and } m \neq n \\ \pi/2, & m, n \text{ integers and } m = n \end{cases}$

25. $\displaystyle\int_0^\pi \sin mx \cos nx \, dx = \begin{cases} 0, & m, n \text{ integers and } m + n \text{ even} \\ 2m/(m^2 - n^2), & m, n \text{ integers and } m + n \text{ odd} \end{cases}$

26. $\displaystyle\int_0^{\pi/2} \sin^2 x \, dx = \int_0^{\pi/2} \cos^2 x \, dx = \frac{\pi}{4}$

27. $\displaystyle\int_0^{\pi/2} \sin^{2m} x \, dx = \int_0^{\pi/2} \cos^{2m} x \, dx = \frac{1 \cdot 3 \cdot 5 \cdots 2m - 1}{2 \cdot 4 \cdot 6 \cdots 2m}\frac{\pi}{2}, \quad m = 1, 2, \cdots$

28. $\displaystyle\int_0^{\pi/2} \sin^{2m+1} x \, dx = \int_0^{\pi/2} \cos^{2m+1} x \, dx = \frac{2 \cdot 4 \cdot 6 \cdots 2m}{1 \cdot 3 \cdot 5 \cdots 2m + 1}, \quad m = 1, 2, \cdots$

29. $\displaystyle\int_0^{\pi/2} \sin^{2p-1} x \cos^{2q-1} x \, dx = \frac{\Gamma(p)\Gamma(q)}{2\Gamma(p+q)}$

30. $\displaystyle\int_0^\infty \frac{\sin px}{x} \, dx = \begin{cases} \pi/2, & p > 0 \\ 0, & p = 0 \\ -\pi/2, & p < 0 \end{cases}$

31. $\displaystyle\int_0^\infty \frac{\sin px \cos qx}{x} \, dx = \begin{cases} 0, & p > q > 0 \\ \pi/2, & 0 < p < q \\ \pi/4, & p = q > 0 \end{cases}$

32. $\displaystyle\int_0^\infty \frac{\sin px \sin qx}{x^2} \, dx = \begin{cases} \pi p/2, & 0 < p \le q \\ \pi q/2, & p \ge q > 0 \end{cases}$

33. $\displaystyle\int_0^\infty \frac{\sin^2 px}{x^2} \, dx = \frac{\pi p}{2}$

34. $\displaystyle\int_0^\infty \frac{x \sin mx}{x^2 + a^2} \, dx = \frac{\pi}{2} e^{-ma}$

35. $\displaystyle\int_0^\infty \frac{1 - \cos px}{x^2} \, dx = \frac{\pi p}{2}$

36. $\displaystyle\int_0^\infty \frac{\sin mx}{x(x^2 + a^2)} \, dx = \frac{\pi}{2a^2}(1 - e^{-ma})$

37. $\displaystyle\int_0^\infty \frac{\cos px - \cos qx}{x} \, dx = \ln\left|\frac{q}{p}\right|$

38. $\displaystyle\int_0^{2\pi} \frac{dx}{a + b \sin x} = \frac{2\pi}{\sqrt{a^2 - b^2}}$

39. $\displaystyle\int_0^\infty \frac{\cos px - \cos qx}{x^2} \, dx = \frac{\pi(q - p)}{2}$

40. $\displaystyle\int_0^{2\pi} \frac{dx}{a + b \cos x} = \frac{2\pi}{\sqrt{a^2 - b^2}}$

41. $\displaystyle\int_0^\infty \frac{\cos mx}{x^2 + a^2} \, dx = \frac{\pi}{2a} e^{-ma}$

42. $\displaystyle\int_0^{\pi/2} \frac{dx}{a + b \cos x} = \frac{\cos^{-1}(b/a)}{\sqrt{a^2 - b^2}}$

43. $\displaystyle\int_0^{2\pi} \frac{dx}{(a + b \sin x)^2} = \int_0^{2\pi} \frac{dx}{(a + b \cos x)^2} = \frac{2\pi a}{(a^2 - b^2)^{3/2}}$

44. $\displaystyle\int_0^{2\pi} \frac{dx}{1 - 2a \cos x + a^2} = \frac{2\pi}{1 - a^2}, \quad 0 < a < 1$

45. $\displaystyle\int_0^\pi \frac{x \sin x \, dx}{1 - 2a \cos x + a^2} = \begin{cases} (\pi/a) \ln|1 + a|, & |a| < 1 \\ \pi \ln|1 + 1/a|, & |a| > 1 \end{cases}$

46. $\displaystyle\int_0^\pi \frac{\cos mx \, dx}{1 - 2a \cos x + a^2} = \frac{\pi a^m}{1 - a^2}, \quad a^2 < 1, \quad m = 0, 1, 2, \cdots$

47. $\displaystyle\int_0^\infty \sin ax^2 \, dx = \int_0^\infty \cos ax^2 \, dx = \frac{1}{2}\sqrt{\frac{\pi}{2a}}$

48. $\displaystyle\int_0^\infty \sin ax^n \, dx = \frac{1}{na^{1/n}}\Gamma(1/n) \sin \frac{\pi}{2n}, \quad n > 1$

49. $\displaystyle\int_0^\infty \cos ax^n \, dx = \frac{1}{na^{1/n}}\Gamma(1/n) \cos \frac{\pi}{2n}, \quad n > 1$

50. $\displaystyle\int_0^\infty \frac{\sin x}{\sqrt{x}} \, dx = \int_0^\infty \frac{\cos x}{\sqrt{x}} \, dx = \sqrt{\frac{\pi}{2}}$

51. $\displaystyle\int_0^\infty \frac{\sin x}{x^p} \, dx = \frac{\pi}{2\Gamma(p) \sin (p\pi/2)}, \quad 0 < p < 1$

52. $\displaystyle\int_0^\infty \frac{\cos x}{x^p} \, dx = \frac{\pi}{2\Gamma(p) \cos (p\pi/2)}, \quad 0 < p < 1$

53. $\displaystyle\int_0^\infty \sin ax^2 \cos 2bx \, dx = \frac{1}{2}\sqrt{\frac{\pi}{2a}}\left(\cos \frac{b^2}{a} - \sin \frac{b^2}{a}\right)$

54. $\displaystyle\int_0^\infty \cos ax^2 \cos 2bx \, dx = \frac{1}{2}\sqrt{\frac{\pi}{2a}}\left(\cos \frac{b^2}{a} + \sin \frac{b^2}{a}\right)$

55. $\displaystyle\int_0^\infty \frac{\sin^3 x}{x^3} \, dx = \frac{3\pi}{8}$

56. $\displaystyle\int_0^\infty \frac{\sin^4 x}{x^4} \, dx = \frac{\pi}{3}$

57. $\displaystyle\int_0^\infty \frac{\tan x}{x} \, dx = \frac{\pi}{2}$

58. $\displaystyle\int_0^{\pi/2} \frac{dx}{1 + \tan^m x} = \frac{\pi}{4}$

59. $\displaystyle\int_0^{\pi/2} \frac{x}{\sin x} \, dx = 2\left\{\frac{1}{1^2} - \frac{1}{3^2} + \frac{1}{5^2} - \frac{1}{7^2} + \cdots\right\}$

60. $\displaystyle\int_0^1 \frac{\tan^{-1} x}{x} \, dx = \frac{1}{1^2} - \frac{1}{3^2} + \frac{1}{5^2} - \frac{1}{7^2} + \cdots$

61. $\displaystyle\int_0^1 \frac{\sin^{-1} x}{x} \, dx = \frac{\pi}{2} \ln 2$

62. $\displaystyle\int_0^1 \frac{1 - \cos x}{x}\, dx \; - \; \int_1^\infty \frac{\cos x}{x}\, dx = \gamma$

63. $\displaystyle\int_0^\infty \left(\frac{1}{1 + x^2} - \cos x \right) \frac{dx}{x} = \gamma$

64. $\displaystyle\int_0^\infty \frac{\tan^{-1} px - \tan^{-1} qx}{x}\, dx = \frac{\pi}{2} \ln\left| \frac{p}{q} \right|$

DEFINITE INTEGRALS INVOLVING EXPONENTIAL FUNCTIONS

65. $\displaystyle\int_0^\infty e^{-ax} \cos bx\, dx = \frac{a}{a^2 + b^2}$

66. $\displaystyle\int_0^\infty e^{-ax} \sin bx\, dx = \frac{b}{a^2 + b^2}$

67. $\displaystyle\int_0^\infty \frac{e^{-ax} \sin bx}{x}\, dx = \tan^{-1}\frac{b}{a}$

68. $\displaystyle\int_0^\infty \frac{e^{-ax} - e^{-bx}}{x}\, dx = \ln\left| \frac{b}{a} \right|$

69. $\displaystyle\int_0^\infty e^{-ax^2}\, dx = \frac{1}{2}\sqrt{\frac{\pi}{a}}$

70. $\displaystyle\int_0^\infty e^{-ax^2} \cos bx\, dx = \frac{1}{2}\sqrt{\frac{\pi}{a}}\, e^{-b^2/4a}$

71. $\displaystyle\int_0^\infty e^{-(ax^2 + bx + c)}\, dx = \frac{1}{2}\sqrt{\frac{\pi}{a}}\, e^{(b^2 - 4ac)/4a}\, \mathrm{erfc}\left(\frac{b}{2\sqrt{a}} \right)$

where $\mathrm{erfc}(p) = \dfrac{2}{\sqrt{\pi}} \displaystyle\int_p^\infty e^{-x^2}\, dx$

72. $\displaystyle\int_0^\infty e^{-(ax^2 + bx + c)}\, dx = \sqrt{\frac{\pi}{a}}\, e^{(b^2 - 4ac)/4a}$

73. $\displaystyle\int_0^\infty x^n e^{-ax}\,dx = \frac{\Gamma(n+1)}{a^{n+1}}$

74. $\displaystyle\int_0^\infty x^m e^{-ax^2}\,dx = \frac{\Gamma[(m+1)/2]}{2a^{(m+1)/2}}$

75. $\displaystyle\int_0^\infty e^{-(ax^2 + b/x^2)}\,dx = \frac{1}{2}\sqrt{\frac{\pi}{a}}\,e^{-2\sqrt{ab}}$

76. $\displaystyle\int_0^\infty \frac{x\,dx}{e^x - 1} = \frac{1}{1^2}\cdot\frac{1}{2^2}\cdot\frac{1}{3^2}\cdot\frac{1}{4^2} + \cdots = \frac{\pi^2}{6}$

77. $\displaystyle\int_0^\infty \frac{x^{n-1}}{e^x - 1}\,dx = \Gamma(n)\left(\frac{1}{1^n} + \frac{1}{2^n} + \frac{1}{3^n} + \cdots\right)$

78. $\displaystyle\int_0^\infty \frac{x\,dx}{e^x + 1} = \frac{1}{1^2} - \frac{1}{2^2} + \frac{1}{3^2} - \frac{1}{4^2} + \cdots = \frac{\pi^2}{12}$

79. $\displaystyle\int_0^\infty \frac{x^{n-1}}{e^x + 1}\,dx = \Gamma(n)\left(\frac{1}{1^n} - \frac{1}{2^n} + \frac{1}{3^n} - \cdots\right)$

80. $\displaystyle\int_0^\infty \frac{\sin mx}{e^{2\pi x} - 1}\,dx = \frac{1}{4}\coth\frac{m}{2} - \frac{1}{2m}$

81. $\displaystyle\int_0^\infty \left(\frac{1}{1+x} - e^{-x}\right)\frac{dx}{x} = \gamma$

82. $\displaystyle\int_0^\infty \frac{e^{-x^2} - e^{-x}}{x}\,dx = \frac{1}{2}\gamma$

83. $\displaystyle\int_0^\infty \left(\frac{1}{e^x - 1} - \frac{e^{-x}}{x}\right)dx = \gamma$

84. $\displaystyle\int_0^\infty \frac{e^{-ax} - e^{-bx}}{x\,\sec px}\,dx = \frac{1}{2}\ln\left(\frac{b^2 + p^2}{a^2 + p^2}\right)$

85. $\displaystyle\int_0^\infty \frac{e^{-ax} - e^{-bx}}{x\,\csc px}\,dx = \tan^{-1}\frac{b}{p} - \tan^{-1}\frac{a}{p}$

86. $\displaystyle\int_0^\infty \frac{e^{-ax}(1-\cos x)}{x^2}\,dx = \cot^{-1} a - \frac{a}{2}\ln(a^2+1)$

DEFINITE INTEGRALS INVOLVING LOGARITHMIC FUNCTIONS

87. $\displaystyle\int_0^1 x^m(\ln|x|)^n\,dx = \frac{(-1)^n n!}{(m+1)^{n+1}}\quad m > -1,\ \ n = 0,1,2,\cdots$

If $n \neq 0,1,2,\cdots$ replace $n!$ by $\Gamma(n+1)$

88. $\displaystyle\int_0^1 \frac{\ln|x|}{1+x}\,dx = -\frac{\pi^2}{12}$

89. $\displaystyle\int_0^1 \frac{\ln|x|}{1-x}\,dx = -\frac{\pi^2}{6}$

90. $\displaystyle\int_0^1 \frac{\ln|1+x|}{x}\,dx = \frac{\pi^2}{12}$

91. $\displaystyle\int_0^1 \frac{\ln|1-x|}{x}\,dx = -\frac{\pi^2}{6}$

92. $\displaystyle\int_0^1 \ln|x|\ln|1+x|\,dx = 2 - 2\ln 2 - \frac{\pi^2}{12}$

93. $\displaystyle\int_0^1 \ln|x|\ln|1-x|\,dx = 2 - \frac{\pi^2}{6}$

94. $\displaystyle\int_0^\infty \frac{x^{p-1}\ln|x|}{1+x}\,dx = -\pi^2 \csc p\pi \cot p\pi \quad 0 < p < 1$

95. $\displaystyle\int_0^1 \frac{x^m - x^n}{\ln|x|}\,dx = \ln\left|\frac{m+1}{n+1}\right|$

96. $\displaystyle\int_0^\infty e^{-x}\ln|x|\,dx = -\gamma$

97. $\displaystyle\int_0^\infty e^{-x^2} \ln|x| \, dx = -\frac{\sqrt{\pi}}{4}(\gamma + 2\ln 2)$

98. $\displaystyle\int_0^\infty \ln\left|\frac{e^x + 1}{e^x - 1}\right| \, dx = \frac{\pi^2}{4}$

99. $\displaystyle\int_0^{\pi/2} \ln|\sin x| \, dx = \int_0^{\pi/2} \ln|\cos x| \, dx = -\frac{\pi}{2}\ln 2$

100. $\displaystyle\int_0^{\pi/2} (\ln|\sin x|)^2 \, dx = \int_0^{\pi/2} (\ln|\cos x|)^2 \, dx = \frac{\pi}{2}(\ln 2)^2 + \frac{\pi^3}{24}$

101. $\displaystyle\int_0^\pi x \ln|\sin x| \, dx = -\frac{\pi^2}{2}\ln 2$

102. $\displaystyle\int_0^{\pi/2} \sin x \ln|\sin x| \, dx = \ln 2 - 1$

103. $\displaystyle\int_0^{2\pi} \ln|a + b\sin x| \, dx = \int_0^{2\pi} \ln|a + b\cos x| \, dx = 2\pi \ln\left|a + \sqrt{a^2 - b^2}\right|$

104. $\displaystyle\int_0^\pi \ln|a + b\cos x| \, dx = \pi \ln\left|\frac{a + \sqrt{a^2 - b^2}}{2}\right|$

105. $\displaystyle\int_0^\pi \ln|a^2 - 2ab\cos x + b^2| \, dx = \begin{cases} 2\pi \ln|a|, & a \geq b > 0 \\ 2\pi \ln|b|, & b \geq a > 0 \end{cases}$

106. $\displaystyle\int_0^{\pi/4} \ln|1 + \tan x| \, dx = \frac{\pi}{8}\ln 2$

107. $\displaystyle\int_0^{\pi/2} \sec x \ln\left|\frac{1 + b\cos x}{1 + a\cos x}\right| \, dx = \frac{1}{2}\{(\cos^{-1} b)^2\}$

108. $\displaystyle\int_0^a \ln\left|2\sin\frac{x}{2}\right| \, dx = -\left(\frac{\sin a}{1^2} + \frac{\sin 2a}{2^2} + \frac{\sin 3a}{3^2} + \cdots\right)$

DEFINITE INTEGRALS INVOLVING HYPERBOLIC FUNCTIONS

109. $\displaystyle\int_0^\infty \frac{\sin ax}{\sinh bx}\,dx = \frac{\pi}{2b}\tanh\frac{a\pi}{2b}$

110. $\displaystyle\int_0^\infty \frac{\cos ax}{\cosh bx}\,dx = \frac{\pi}{2b}\operatorname{sech}\frac{a\pi}{2b}$

111. $\displaystyle\int_0^\infty \frac{x\,dx}{\sinh ax} = \frac{\pi^2}{4a^2}$

112. $\displaystyle\int_0^\infty \frac{x^n\,dx}{\sinh bx} = \frac{2^{n+1}-1}{2^n a^{n+1}}\Gamma(n+1)\left\{\frac{1}{1^{n+1}}+\frac{1}{2^{n+1}}+\frac{1}{3^{n+1}}+\cdots\right\}$

113. $\displaystyle\int_0^\infty \frac{\sinh ax}{e^{bx}+1}\,dx = \frac{\pi}{2b}\csc\frac{a\pi}{b}-\frac{1}{2a}$

114. $\displaystyle\int_0^\infty \frac{\sinh ax}{e^{bx}-1}\,dx = \frac{1}{2a}-\frac{\pi}{2b}\cot\frac{a\pi}{b}$

MISCELLANEOUS DEFINITE INTEGRALS

115. $\displaystyle\int_0^\infty \frac{f(ax)-f(bx)}{x}\,dx = \{f(0)-f(\infty)\}\ln\left|\frac{b}{a}\right|$

This is called *Frullani's integral*. It holds if $f'(x)$ is continuous and $\displaystyle\int_0^\infty \frac{f(x)-f(\infty)}{x}\,dx$ converges.

116. $\displaystyle\int_0^1 \frac{dx}{x^x} = \frac{1}{1^1}+\frac{1}{2^2}+\frac{1}{3^3}+\cdots$

117. $\displaystyle\int_{-a}^a (a+x)^{m-1}(a-x)^{n-1}\,dx = (2a)^{m+n-1}\frac{\Gamma(m)\Gamma(n)}{\Gamma(m+n)}$

CHAPTER 11

Series

11.1

Series of Constants

1. $1 + 2 + 3 + \cdots + n = \frac{1}{2}n(n + 1)$

2. $1 + 3 + 5 + \cdots + (2n - 1) = n^2$

3. $1^2 + 2^2 + 3^2 + \cdots + n^2 = \dfrac{n(n + 1)(2n + 1)}{6}$

4. $1^3 + 2^3 + 3^3 + \cdots + n^3 = \dfrac{n^2(n + 1)^2}{4} = (1 + 2 + 3 + \cdots + n)^2$

5. $1^4 + 2^4 + 3^4 + \cdots + n^4 = \dfrac{n(n + 1)(2n + 1)(3n^2 + 3n - 1)}{30}$

6. $1^5 + 2^5 + 3^5 + \cdots + n^5 = \dfrac{n^6}{6} + \dfrac{n^5}{2} + \dfrac{5n^4}{12} - \dfrac{n^2}{12}$

7. $1^2 + 3^2 + 5^2 + \cdots + (2n - 1)^2 = \frac{1}{3}n(4n^2 - 1)$

8. $1^3 + 3^3 + 5^3 + \cdots + (2n - 1)^3 = n^2(2n^2 - 1)$

9. $2^2 + 4^2 + 6^2 + 8^2 + \cdots + (2n)^2 = \dfrac{2n(n + 1)(2n + 1)}{3}$

10. $1 \cdot 2 + 2 \cdot 3 + 3 \cdot 4 + \cdots + n(n + 1) = \dfrac{n(n + 1)(n + 2)}{3}$

11. $1 \cdot 2 \cdot 3 + 2 \cdot 3 \cdot 4 + 3 \cdot 4 \cdot 5 + \cdots + n(n + 1)(n + 2) = \frac{1}{4}n(n + 1)(n + 2)(n + 3)$

12. $1 - \frac{1}{2} + \frac{1}{3} - \frac{1}{4} + \frac{1}{5} - \cdots = \ln 2$

13. $1 - \frac{1}{3} + \frac{1}{5} - \frac{1}{7} + \frac{1}{9} - \cdots = \frac{\pi}{4}$

14. $\dfrac{1}{1 \cdot 3} + \dfrac{1}{3 \cdot 5} + \dfrac{1}{5 \cdot 7} + \dfrac{1}{7 \cdot 9} + \cdots = \frac{1}{2}$

15. $\dfrac{1}{1 \cdot 3} + \dfrac{1}{2 \cdot 4} + \dfrac{1}{3 \cdot 5} + \dfrac{1}{4 \cdot 6} + \cdots = \frac{3}{4}$

11.2

Taylor Series

TAYLOR SERIES FOR FUNCTIONS OF ONE VARIABLE

15. $f(x) = f(c) + f'(c)(x - c) + \dfrac{f''(c)(x - c)^2}{2!} + \cdots + \dfrac{f^{(n-1)}(c)(x - c)^{n-1}}{(n-1)!} + R_n$

where R_n, the remainder after n terms, is given by either of the following forms:

Lagrange's form $\quad R_n = \dfrac{f^{(n)}(z)(x - a)^n}{n!}$

Cauchy's form $\quad R_n = \dfrac{f^{(n)}(z)(x - z)^{n-1}(x - a)}{(n - 1)!}$

The value z, which may be different in the two forms, lies between c and x. The result holds if $f(x)$ has continuous derivatives of order n at least.

If $\lim\limits_{n \to \infty} R_n = 0$, the infinite series obtained is called the *Taylor Series* for $f(x)$ about $x = a$. If $c = 0$ the series is often called a *Maclaurin series*. These series, often called power series, generally converge for all values of x in some interval called the *interval of convergence* and diverge for all x outside this interval.

BINOMIAL SERIES

16. $(a + x)^n = a^n + na^{n-1}x + \dfrac{n(n-1)}{2!}a^{n-2}x^2 + \dfrac{n(n-1)(n-2)}{3!}a^{n-3}x^3 + \cdots$

$= a^n + \dbinom{n}{1}a^{n-1}x + \dbinom{n}{2}a^{n-2}x^2 + \dbinom{n}{3}a^{n-3}x^3 + \cdots$

17. $(a + x)^2 = a^2 + 2ax + x^2$

18. $(a + x)^3 = a^3 + 3a^2x + 3ax^2 + x^3$

19. $(a + x)^4 = a^4 + 4a^3x + 6a^2x^2 + 4ax^3 + x^4$

20. $(1 + x)^{-1} = 1 - x + x^2 - x^3 + x^4 - \cdots$ $\qquad\qquad -1 < x < 1$

21. $(1 + x)^{-2} = 1 - 2x + 3x^2 - 4x^3 + 5x^4 - \cdots$ $\qquad -1 < x < 1$

22. $(1 + x)^{-3} = 1 - 3x + 6x^2 - 10x^3 + 15x^4 - \cdots$ $\qquad -1 < x < 1$

23. $(1 + x)^{-1/2} = 1 - \dfrac{1}{2}x + \dfrac{1 \cdot 3}{2 \cdot 4}x^2 - \dfrac{1 \cdot 3 \cdot 5}{2 \cdot 4 \cdot 6}x^3 + \cdots$ $\qquad -1 < x \leq 1$

24. $(1 + x)^{1/2} = 1 + \dfrac{1}{2}x - \dfrac{1}{2 \cdot 4}x^2 + \dfrac{1 \cdot 3}{2 \cdot 4 \cdot 6}x^3 - \cdots$ $\qquad -1 < x \leq 1$

25. $(1 + x)^{-1/3} = 1 - \dfrac{1}{3}x - \dfrac{1 \cdot 4}{3 \cdot 6}x^2 - \dfrac{1 \cdot 4 \cdot 7}{3 \cdot 6 \cdot 9}x^3 + \cdots$ $\qquad -1 < x \leq 1$

26. $(1 + x)^{1/3} = 1 + \dfrac{1}{3}x - \dfrac{2}{3 \cdot 6}x^2 + \dfrac{2 \cdot 5}{3 \cdot 6 \cdot 9}x^3 - \cdots$ $\qquad -1 < x \leq 1$

SERIES FOR EXPONENTIAL AND LOGARITHMIC FUNCTIONS

27. $e^x = 1 + x + \dfrac{x^2}{2!} + \dfrac{x^3}{3!} + \cdots$ $\qquad\qquad -\infty < x < \infty$

28. $a^x = e^{x \ln|a|} = 1 + x \ln|a| + \dfrac{(x \ln|a|)^2}{2!} + \dfrac{(x \ln|a|)^3}{3!} + \cdots$ $\quad -\infty < x < \infty$

29. $\ln|1 + x| = x - \dfrac{x^2}{2} + \dfrac{x^3}{3} - \dfrac{x^4}{4} + \cdots$ $\qquad\qquad -1 < x \le 1$

30. $\dfrac{1}{2} \ln\left|\dfrac{1+x}{1-x}\right| = x + \dfrac{x^3}{3} + \dfrac{x^5}{5} + \dfrac{x^7}{7} + \cdots$ $\qquad -1 < x < 1$

31. $\ln|x| = 2\left\{\left(\dfrac{x-1}{x+1}\right) + \dfrac{1}{3}\left(\dfrac{x-1}{x+1}\right)^3 + \dfrac{1}{5}\left(\dfrac{x-1}{x+1}\right)^5 + \cdots\right\}$ $\qquad x > 0$

32. $\ln|x| = \left(\dfrac{x-1}{x}\right) + \dfrac{1}{2}\left(\dfrac{x-1}{x}\right)^2 + \dfrac{1}{3}\left(\dfrac{x-1}{x}\right)^3 + \cdots$ $\qquad x \ge \dfrac{1}{2}$

SERIES FOR TRIGONOMETRIC FUNCTIONS

33. $\sin x = x - \dfrac{x^3}{3!} + \dfrac{x^5}{5!} - \dfrac{x^7}{7!} + \cdots$ $\qquad\qquad -\infty < x < \infty$

34. $\cos x = 1 - \dfrac{x^2}{2!} + \dfrac{x^4}{4!} - \dfrac{x^6}{6!} + \cdots$ $\qquad\qquad -\infty < x < \infty$

35. $\tan x = x + \dfrac{x^3}{3} + \dfrac{2x^5}{15} + \dfrac{17x^7}{315} + \cdots + \dfrac{2^{2n}(2^{2n} - 1)B_n x^{2n-1}}{(2n)!} + \cdots$ $\qquad |x| < \dfrac{\pi}{2}$

36. $\cot x = \dfrac{1}{x} - \dfrac{x}{3} - \dfrac{x^3}{45} - \cdots - \dfrac{2^{2n} B_n x^{2n-1}}{(2n)!} - \cdots$ $\qquad 0 < |x| < \pi$

37. $\sec x = 1 + \dfrac{x^2}{2} + \dfrac{5x^4}{24} + \dfrac{61x^6}{720} + \cdots + \dfrac{E_n x^{2n}}{(2n)!} + \cdots$ $\qquad |x| < \dfrac{\pi}{2}$

38. $\csc x = \dfrac{1}{x} + \dfrac{x}{6} + \dfrac{7x^3}{360} + \dfrac{31x^5}{15{,}120} + \cdots + \dfrac{2(2^{2n-1} - 1)B_n x^{2n-1}}{(2n)!} + \cdots$ $\qquad 0 < |x| < \pi$

39. $\sin^{-1} x = x + \dfrac{1}{2}\dfrac{x^3}{3} + \dfrac{1 \cdot 3}{2 \cdot 4}\dfrac{x^5}{5} + \dfrac{1 \cdot 3 \cdot 5}{2 \cdot 4 \cdot 6}\dfrac{x^7}{7} + \cdots$ $\qquad |x| < 1$

40. $\cos^{-1} x = \dfrac{\pi}{2} - \sin^{-1} x = \dfrac{\pi}{2} - \left(x + \dfrac{1}{2}\dfrac{x^3}{3} + \dfrac{1 \cdot 3}{2 \cdot 4}\dfrac{x^5}{5} + \cdots\right)$ $\qquad |x| < 1$

41. $\tan^{-1} x = \begin{cases} x - \dfrac{x^3}{3} + \dfrac{x^5}{5} - \dfrac{x^7}{7} + \cdots, & \text{if } x^2 < 1 \quad \text{or} \\[2mm] \dfrac{\pi}{2} - \dfrac{1}{x} + \dfrac{1}{3x^3} - \dfrac{1}{5x^5} + \cdots, & \text{if } x \ge 1 \\[2mm] -\dfrac{\pi}{2} - \dfrac{1}{x} + \dfrac{1}{3x^3} - \dfrac{1}{5x^5} + \cdots, & \text{if } x \le 1 \end{cases}$

42. $\cot^{-1} x = \dfrac{\pi}{2} - \tan^{-1} x = \begin{cases} \dfrac{\pi}{2} - \left(x - \dfrac{x^3}{3} + \dfrac{x^5}{5} - \cdots\right), & \text{if } x^2 < 1 \quad \text{or} \\[2mm] p\pi + \dfrac{1}{x} - \dfrac{1}{3x^3} + \dfrac{1}{5x^5} - \cdots, & [p = 0 \text{ if } x > 1, \ p = 1 \text{ if } x < -1 \end{cases}$

43. $\sec^{-1} x = \cos^{-1} \dfrac{1}{x} = \dfrac{\pi}{2} - \left(\dfrac{1}{x} + \dfrac{1}{2 \cdot 3x^3} + \dfrac{1 \cdot 3}{2 \cdot 4 \cdot 5x^5} + \cdots \right)$　　　　$|x| > 1$

44. $\csc^{-1} x = \sin^{-1} \dfrac{1}{x} = \dfrac{1}{x} + \dfrac{1}{2 \cdot 3x^3} + \dfrac{1 \cdot 3}{2 \cdot 4 \cdot 5x^5} + \cdots$　　　　$|x| > 1$

SERIES FOR HYPERBOLIC FUNCTIONS

45. $\sinh x = x + \dfrac{x^3}{3!} + \dfrac{x^5}{5!} + \dfrac{x^7}{7!} + \cdots$　　　　　　　　$-\infty < x < \infty$

46. $\cosh x = 1 + \dfrac{x^2}{2!} + \dfrac{x^4}{4!} + \dfrac{x^6}{6!} + \cdots$　　　　　　　　$-\infty < x < \infty$

47. $\tanh x = x - \dfrac{x^3}{3} + \dfrac{2x^5}{15} + \dfrac{17x^7}{315} + \cdots \dfrac{(-1)^{n-1} 2^{2n}(2^{2n} - 1) B_n x^{2n-1}}{(2n)!} + \cdots$　　$|x| < \dfrac{\pi}{2}$

48. $\coth x = \dfrac{1}{x} + \dfrac{x}{3} - \dfrac{x^3}{45} + \dfrac{2x^5}{945} + \cdots \dfrac{(-1)^{n-1} 2^{2n} B_n x^{2n-1}}{(2n)!} + \cdots$　　$0 < |x| < \pi$

49. $\operatorname{sech} x = 1 - \dfrac{x^2}{2} + \dfrac{5x^4}{24} - \dfrac{61x^6}{720} + \cdots \dfrac{(-1)^n E_n x^{2n}}{(2n)!} + \cdots$　　$|x| < \dfrac{\pi}{2}$

50. $\operatorname{csch} x = \dfrac{1}{x} - \dfrac{x}{6} + \dfrac{7x^3}{360} - \dfrac{31x^5}{15,120} + \cdots \dfrac{(-1)^n 2(2^{2n-1} - 1) B_n x^{2n-1}}{(2n)!} + \cdots$　　$0 < |x| < \pi$

51. $\sinh^{-1} x = \begin{cases} x - \dfrac{x^3}{2 \cdot 3} + \dfrac{1 \cdot 3 x^5}{2 \cdot 4 \cdot 5} - \dfrac{1 \cdot 3 \cdot 5 x^7}{2 \cdot 4 \cdot 6 \cdot 7} + \cdots, \ |x| < 1 \quad \text{or} \\[2mm] \pm \left(\ln|2x| + \dfrac{1}{2 \cdot 2x^2} - \dfrac{1 \cdot 3}{2 \cdot 4 \cdot 4x^4} + \dfrac{1 \cdot 3 \cdot 5}{2 \cdot 4 \cdot 6 \cdot 6x^6} - \cdots \right) \end{cases}$ $\begin{bmatrix} \text{``+''} \ \text{if} \ x \geq 1 \\ \text{``-''} \ \text{if} \ x \leq -1 \end{bmatrix}$

52. $\cosh^{-1} x = \pm \left\{ \ln|2x| - \left(\dfrac{1}{2 \cdot 2x^2} + \dfrac{1 \cdot 3}{2 \cdot 4 \cdot 4x^4} + \dfrac{1 \cdot 3 \cdot 5}{2 \cdot 4 \cdot 6 \cdot 6x^6} + \cdots \right) \right\}$ $\begin{bmatrix} \text{``+''} \ \text{if} \ \cosh^{-1} x > 0, \ x \geq 1 \\ \text{``-''} \ \text{if} \ \cosh^{-1} x < 0, \ x \geq 1 \end{bmatrix}$

53. $\tanh^{-1} x = x + \dfrac{x^3}{3} + \dfrac{x^5}{5} + \dfrac{x^7}{7} + \cdots$　　　　　　　$|x| < 1$

54. $\coth^{-1} x = \dfrac{1}{x} + \dfrac{1}{3x^3} + \dfrac{1}{5x^5} + \dfrac{1}{7x^7} + \cdots$　　　　　　　$|x| > 1$

MISCELLANEOUS SERIES

55. $e^{\sin x} = 1 + x + \dfrac{x^2}{2} - \dfrac{x^4}{8} - \dfrac{x^5}{15} + \cdots$　　　　　　　$-\infty < x < \infty$

56. $e^{\cos x} = e \left(1 - \dfrac{x^2}{2} + \dfrac{x^4}{6} - \dfrac{31x^6}{720} + \cdots \right)$　　　　　$-\infty < x < \infty$

57. $e^{\tan x} = 1 + x + \dfrac{x^2}{2} + \dfrac{x^3}{2} + \dfrac{3x^4}{8} + \cdots$　　　　　　$|x| < \dfrac{\pi}{2}$

58. $e^x \sin x = x + x^2 + \dfrac{2x^3}{3} - \dfrac{x^5}{30} - \dfrac{x^6}{90} + \cdots + \dfrac{2^{n/2} \sin(n\pi/4) x^n}{n!} + \cdots$　　$-\infty < x < \infty$

59. $e^x \cos x = 1 + x - \dfrac{x^3}{3} - \dfrac{x^4}{6} + \cdots + \dfrac{2^{n/2} \cos(n\pi/4) x^n}{n!} + \cdots$　　$-\infty < x < \infty$

60. $\ln|\sin x| = \ln|x| - \dfrac{x^2}{6} - \dfrac{x^4}{180} - \dfrac{x^6}{2835} - \cdots - \dfrac{2^{2n-1} B_n x^{2n}}{n(2n)!} + \cdots$　　$0 < |x| < \pi$

61. $\ln|\cos x| = -\dfrac{x^2}{2} - \dfrac{x^4}{12} - \dfrac{x^6}{45} - \dfrac{17x^8}{2520} - \cdots - \dfrac{2^{2n-1}(2^{2n}-1)B_n x^{2n}}{n(2n)!} + \cdots \qquad |x| < \dfrac{\pi}{2}$

62. $\ln|\tan x| = \ln|x| + \dfrac{x^2}{3} + \dfrac{7x^4}{90} + \dfrac{62x^6}{2835} + \cdots + \dfrac{2^{2n}(2^{2n-1}-1)B_n x^{2n}}{n(2n)!} + \cdots \qquad 0 < |x| < \dfrac{\pi}{2}$

63. $\dfrac{\ln|1+x|}{1+x} = x - (1+\tfrac{1}{2})x^2 + (1+\tfrac{1}{2}+\tfrac{1}{3})x^3 - \cdots \qquad |x| < 1$

APPENDICES

A.

Mathematical Symbols

$\pm$	Plus or minus	$\mp$	Minus or plus
$=$	Equal to	$\neq$	Not equal to
$\equiv$	Identically equal	$\not\equiv$	Not identically equal to
$\times$	Multiplication	$\div, /$	Division
$>$	Greater than	$<$	Less than
$\geq$	Greater than or equal to	$\leq$	Less than or equal to
$\simeq$	Congruent to	$\approx$	Approximately equal to
$\sim$	Similar to	$\rightarrow$	Approaches
$\propto$	Varies	$\ln$	Natural logarithm
$\perp$	Perpendicular to	$\parallel$	Parallel to
$\sqrt{}$	Square root	$\sqrt[n]{}$	nth root
i	Imaginary unit $\sqrt{-1}$	∞	Infinity
${}_nC_r$ or $\binom{n}{r}$ $\dfrac{n!}{r!(n-r)!}$		${}_nP_r$ $n!/(n-r)!$	
$\lvert a \rvert$	Absolute value of a	$\cdots$	And so on

B.

Greek Alphabet

A	α	Alpha	I	ι	Iota	P	ρ	Rho
B	β	Beta	K	κ	Kappa	Σ	σ	Sigma
Γ	γ	Gamma	Λ	λ	Lambda	T	τ	Tau
Δ	δ	Delta	M	μ	Mu	Υ	υ	Upsilon
E	ϵ	Epsilon	N	ν	Nu	Φ	ϕ	Phi
Z	ζ	Zeta	Ξ	ξ	Xi	X	χ	Chi
H	η	Eta	O	o	Omicron	Ψ	ψ	Psi
Θ	θ	Theta	Π	π	Pi	Ω	ω	Omega

C.

Answers

Problem Set 1, Page 11

1. a. $|\overline{TI}| = |\overline{AG}|, |\overline{RI}| = |\overline{NG}|$ **b.** $|\overline{RI}| = |\overline{NG}|$

    c. $\angle T \simeq \angle A$

2. a. $|\overline{CB}| = |\overline{FE}|, |\overline{AB}| = |\overline{DE}|$

    b. $|\overline{AB}| = |\overline{DE}|$ **c.** $\angle C \simeq \angle F$

3. $\overline{AB} \simeq \overline{ED}; \overline{AC} \simeq \overline{EF}; \overline{CB} \simeq \overline{FD}; \angle A \simeq \angle E;$

    $\angle B \simeq \angle D; \angle C \simeq \angle F$

4. $\overline{GH} \simeq \overline{G'H'}; \overline{GI} \simeq \overline{G'I'}; \overline{HI} \simeq \overline{H'I'};$

    $\angle G \simeq \angle G'; \angle H \simeq \angle H'; \angle I \simeq \angle I'$

5. $\overline{RS} \simeq \overline{TU}; \overline{RT} \simeq \overline{RT}; \overline{ST} \simeq \overline{UR};$

    $\angle SRT \simeq \angle UTR; \angle S \simeq \angle U;$

    $\angle STR \simeq \angle URT$

6. $\overline{WX} \simeq \overline{YZ}; \overline{WZ} \simeq \overline{YX}; \overline{XZ} \simeq \overline{XZ};$

    $\angle W \simeq \angle Y; \angle WXZ \simeq \angle YZX;$

    $\angle WZX \simeq \angle YXZ$

7. Congruent, ASA

8. Congruent, SAS (or possibly SSS)

9. Congruent, SAS

10. Congruent, ASA

11. similar   **12.** not similar

13. not similar   **14.** similar

15. similar   **16.** similar

17. 4   **18.** 6

19. $\frac{16}{3}$ **20.** $\frac{40}{7}$ **21.** $\frac{8}{3}$ **22.** $\frac{10}{9}$

23. 24 ft   **24.** $\sqrt{640} \approx 25$ ft

25. 10 ft   **26.** $2\sqrt{2}$ **27.** $3\sqrt{2}$ **28.** 29 ft

29. $5\sqrt{13}$ ft; 18 ft; 55 ft

30. $\sqrt{3}$ **31.** $\frac{20}{3}$ **32.** 260 ft   **33.** 125

34. 24 ft   **35.** 29 ft   **36.** 9.19 cm^3 **37.** 4 ft

38. 3 ft   **39.** $\frac{32}{3}\pi$

Problem Set 2, Page 35

1. 2^{12} **2.** 5^{10} **3.** 5^6 **4.** 2^{24}

5. 3^3 **6.** 3^{-3} **7.** 2^3 **8.** 2^{-13}

9. 2   **10.** 2^3 **11.** 2^2 **12.** 2

13. $a^5 b^8 c^7$ **14.** $b^{-1}c^3$ **15.** $a^4 b^8 c^{12}$

16. $a^{10} b^{15} c^{-10}$ **17.** $a^2 b^{-2} c^2$ **18.** $a^2 b^2 c^2$

19. $a^4 b^4 c^{-4}$ **20.** $\dfrac{x^2 + y^2}{x^2 y^2}$

21. $\dfrac{x^4 y^4}{(x^2 + y^2)^2}$ **22.** $a^{15/2} b^{10} c^3$

23. $a^{-1/2} b^{-1/2}(2a + 3b)$

24. $a^{-1/2}(4a - 3b)$ **25.** $a^{-1/2} b(a + b)$

26. $a^{-1/2} b^{-1/3}(a + b^{2/3})$

27. $(x + 1)^2 (2x + 3)(10x + 13)$

28. $2x(x^2 - 2)^4 (2x^2 + 1)^3 (18x^2 - 11)$

29. $[7, \infty)$ **30.** $(-\infty, 36)$

31. $[\frac{1}{2}, \infty)$ **32.** $(-\infty, 2)$

33. $[-1, 5]$ **34.** $(-8, -3]$

35. $(-\frac{7}{3}, 1]$ **36.** $(-\frac{15}{2}, 4]$

37. $\{1, -6\}$ **38.** $\{-2, -3)$

39. $\{0, \frac{7}{3}\}$ **40.** $\left\{\pm\sqrt{\frac{2}{7}}\right\}$

41. $\left\{\dfrac{-5 \pm \sqrt{73}}{8}\right\}$ **42.** $\left\{\dfrac{3}{2}\right\}$

43. $\left\{\dfrac{-1 \pm \sqrt{8w + 1}}{4}\right\}$ **44.** $\left\{\dfrac{-w \pm \sqrt{w^2 - 40}}{4}\right\}$

45. $\left\{\dfrac{1 \pm i}{2}\right\}$ **46.** $\left\{\dfrac{-1 \pm \sqrt{8y - 47}}{4}\right\}$

47. $\left\{2, \dfrac{3t + 2}{4}\right\}$ **48.** $x = 3 \pm \sqrt{4 - (y - 2)^2}$

49. $(-\infty, -\frac{5}{2}) \cup (-1, \frac{7}{3})$ **50.** $(-\frac{2}{3}, \frac{3}{2}) \cup (2, \infty)$

51. $(-\infty, 0) \cup (\frac{1}{2}, 5)$ **52.** $(-\infty, -\frac{3}{2}) \cup (0, 2)$

53. $(-\infty, \infty)$ **54.** $[1 - \sqrt{7}, 1 + \sqrt{7}]$

55. -6 **56.** 1   **57.** -9 **58.** 7

59. 2   **60.** 1   **61.** -3

62. $-17\mathbf{i} + 8\mathbf{j} + 14\mathbf{k}$ **63.** -1 **64.** 2

65. 0   **66.** 0

67. $a^4 + 4a^3 b + 6a^2 b^2 + 4ab^3 + b^4$

68. $a^3 + 6a^2 b + 12ab^2 + 8b^3$

69. $27x^3 + 54x^2 y + 36xy^2 + 8y^3$

70. $-x^5 + 10x^4 y - 40x^3 y^2 + 80x^2 y^3$
$- 80xy^4 + 32y^5$

71. $\dfrac{1}{x^4} + \dfrac{4y}{x^3} + \dfrac{6y^2}{x^2} + \dfrac{4y^3}{x} + y^4$

72. $a^{5/2} + 10a^2 b + 40b^2 a^{3/2} + 80ab^3$
$+ 80a^{1/2} b^4 + 32b^5$

73. $\{1, 2, -2\}$ **74.** $\{3, -3, \frac{1}{2}\}$

75. $\{2, 3, -3\}$ **76.** $\{-1, 2, -3\}$

77. $\{2, -2, -3\}$ **78.** $\{2, -3, \frac{1}{2}\}$

79. $\{-3, 5, -\frac{1}{2}\}$ **80.** $\{3\}$

81. $\{1, -1, 3, -6\}$ **82.** $\{2, -2, 3, -3\}$

83. $\{-3, -5, -7\}$

84. $\{-1, -2, 3, -4\}$ **85.** $\{1, -1\}$

86. $\{0, -1, -3, 1 - \sqrt{2}, 1 + \sqrt{2}\}$

Problem Set 3, Page 49

1. $\dfrac{\pi}{30}$ **2.** $\dfrac{3\pi}{2}$ **3.** $\dfrac{5\pi}{9}$ **4.** $\dfrac{\pi^2}{90}$

5. $120°$ **6.** $45n°$ **7.** $\left(\dfrac{756}{\pi}\right)° \approx 240.64°$

8. $\left(\dfrac{21{,}600}{\pi}\right)° \approx 6{,}875.49°$ (This is about 19 revolutions.)

9. a. -1 **b.** 0 **c.** 1 **d.** 0 **e.** -1

10. a. $\dfrac{\sqrt{2}}{2}$ **b.** $-\dfrac{\sqrt{2}}{2}$ **c.** $-\dfrac{\sqrt{2}}{2}$ **d.** $-\dfrac{\sqrt{2}}{2}$
e. 0

11. 1

12.

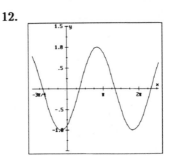

13.

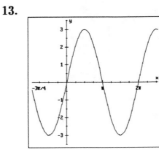

14.

15.

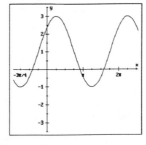

16.

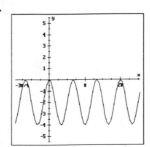

17.

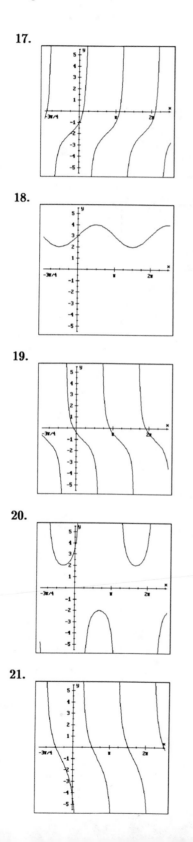

18.

19.

20.

21.

22. $\dfrac{\sqrt{6} - \sqrt{2}}{4}$ **23.** $\dfrac{\sqrt{6} - \sqrt{2}}{4}$

24. $\dfrac{\sqrt{2} + \sqrt{6}}{4}$ **25.** $\dfrac{\sqrt{2} - \sqrt{6}}{4}$

26. $\sin 2x = \sin(x + x)$

$\qquad = \sin x \cos x + \cos x \sin x = 2 \sin x \cos x$

27. $\sin(x + n\pi) = \sin x \cos n\pi + \cos x \sin n\pi$

$\qquad = (-1)^{n}\sin x$

since $\sin n\pi = 0$ and $\cos n\pi = (-1)^{n}$

28. $\cos(x + n\pi) = \cos x \cos n\pi - \sin x \sin n\pi$

$\qquad = (-1)^{n}\cos x$

since $\sin n\pi = 0$ and $\cos n\pi = (-1)^{n}$

29. $\quad \sin(x + y) = \sin x \cos y + \cos x \sin y$

$\qquad \sin(x - y) = \sin x \cos y - \cos x \sin y$

Thus,

$\qquad \sin(x + y) + \sin(x - y) = 2 \sin x \cos y$

30. From the work in Problem 29,

$\qquad \sin(x + y) - \sin(x - y) = 2 \cos x \sin y$

31. Let $A = x + y$ and $B = x - y$, then
$x = \dfrac{A + B}{2}$ and $y = \dfrac{A - B}{2}$. Then, by
substitution into Problem 29, we have
$\sin A + \cos B = 2 \sin\left(\dfrac{A + B}{2}\right)\cos\left(\dfrac{A - B}{2}\right)$

32. Let $\alpha = \sin^{-1}\frac{1}{5}$ so $\sin \alpha = \frac{1}{5}$; and

$\beta = \cos^{-1}\frac{1}{5}$ so $\cos \beta = \frac{1}{5}$

Then $\cos \alpha = \sqrt{1 - \left(\frac{1}{5}\right)^{2}} = \frac{2}{5}\sqrt{6}$

and $\sin \beta = \frac{2}{5}\sqrt{6}$.

$\cos(\alpha + 2\beta) = \cos \alpha \cos 2\beta - \sin \alpha \sin 2\beta$

$\qquad = \cos \alpha(\cos^{2}\beta - \sin^{2}\beta) - \sin \alpha(2 \sin \beta \cos \beta)$

$\qquad = -\frac{2}{5}\sqrt{6}$

33. Let $\alpha = \sin^{-1}\frac{1}{5}$ so $\sin \alpha = \frac{1}{5}$; and

$\beta = \cos^{-1}\frac{1}{4}$ so $\cos \beta = \frac{1}{4}$

Then $\cos \alpha = \frac{2}{5}\sqrt{6}$ and

$\sin \beta = \sqrt{1 - \left(\frac{1}{4}\right)^{2}} = \frac{1}{4}\sqrt{15}.$

$$\sin(\alpha + \beta) = \sin \alpha \cos \beta + \cos \alpha \sin \beta$$

$$= \tfrac{1}{5} \cdot \tfrac{1}{4} + \tfrac{2}{5}\sqrt{6} \cdot \tfrac{1}{4}\sqrt{15} = \frac{1 + 6\sqrt{10}}{20}$$

34. $\dfrac{\pi}{6}, \dfrac{5\pi}{6}$ **35.** $\dfrac{7\pi}{6}, \dfrac{11\pi}{6}$

36. $0, \dfrac{\pi}{2}, \pi, \dfrac{3\pi}{2}$ **37.** $0, \pi$

38. $\dfrac{\pi}{3}, \dfrac{3\pi}{4}, \dfrac{5\pi}{4}, \dfrac{5\pi}{3}$ **39.** $\dfrac{\pi}{6}, \dfrac{5\pi}{6}, \dfrac{7\pi}{6}, \dfrac{11\pi}{6}$

40. $0, \dfrac{\pi}{4}, \pi, \dfrac{5\pi}{4}$ **41.** $0, \dfrac{\pi}{3}, \pi, \dfrac{4\pi}{3}$

42. 2.24, 4.05   **43.** 4.71

44. 0.36, 1.21, 3.51, 4.35   **45.** 0.67, 2.48

46. 0.52, 1.57, 1.83, 2.62, 2.88, 3.67, 4.71, 4.97,

5.76, 6.02

47. 0.00, 1.57, 2.09, 3.14, 4.19, 4.71

48. 0.68, 2.08, 2.78, 4.17, 4.87, 6.27

49. 0.02, 1.59, 3.17, 4.74

50. 0.41, 1.16, 3.55, 4.30

51. 0.00, 1.05, 1.22, 1.92, 2.09, 3.14, 3.32, 4.01,

4.19, 5.24, 5.41, 6.11

52. 0.00, 1.05, 1.22, 1.92, 2.09, 3.14, 3.32, 4.01,

4.19, 5.24, 5.41, 6.11

Problem Set 4, Page 59

1. $V(0, 0); c = 2$ **2.** $V(0, 0); c = 3$

3. $V(0, 0); c = 5$ **4.** $V(0, 0); c = 5/8$

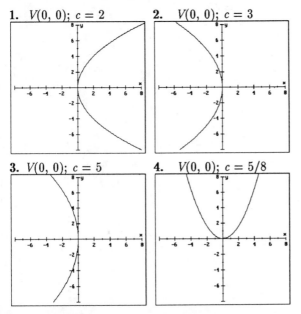

5. $V(0, 0); c = 1$ **6.** $V(0, 0); c = 1/2$

7. $V(0, 0); c = 5/8$ **8.** $V(0, 0); c = 3/4$

9. $V(0, 0); c = 5/4$ **10.** $V(4, 0); c = 3/16$

11. $x^2 = -\dfrac{4}{5}(y - 5)$ **12.** $x^2 = -3(y - 4)$

$V(0, 5); c = 1/5$ $V(0, 4); c = 3/4$

13. $V(-2, 1); c = 1/2$ **14.** $V(1, -3); c = 3/4$

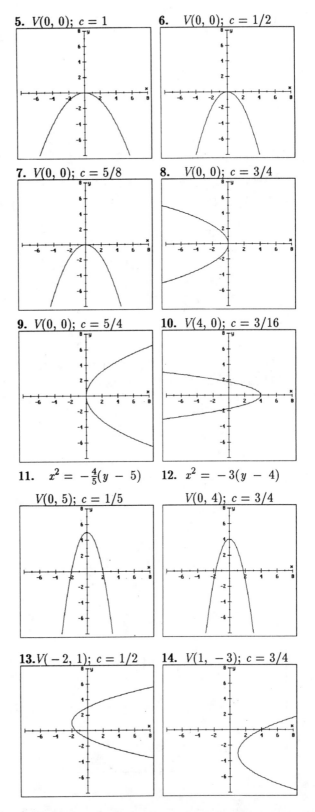

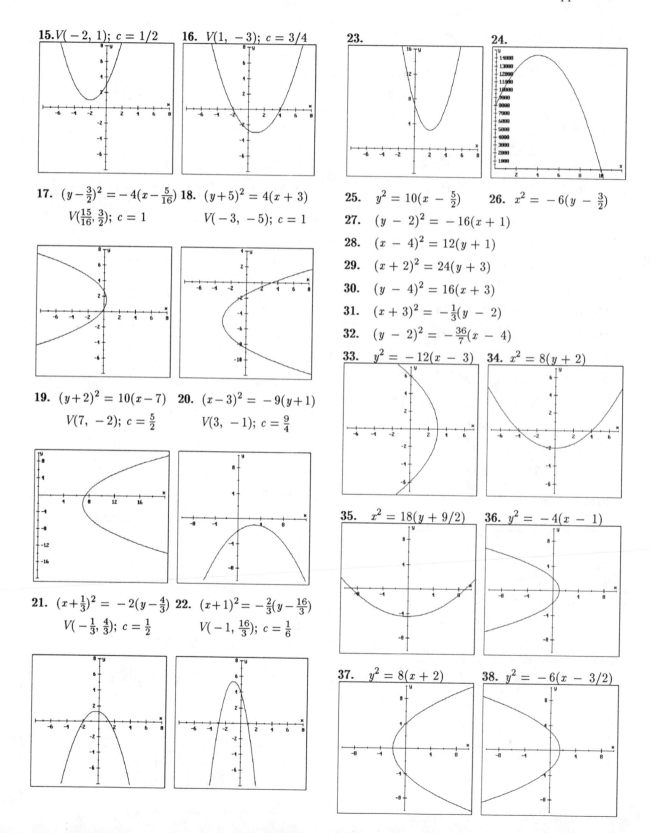

15. $V(-2, 1)$; $c = 1/2$

16. $V(1, -3)$; $c = 3/4$

17. $(y - \frac{3}{2})^2 = -4(x - \frac{5}{16})$

$V(\frac{15}{16}, \frac{3}{2})$; $c = 1$

18. $(y + 5)^2 = 4(x + 3)$

$V(-3, -5)$; $c = 1$

19. $(y + 2)^2 = 10(x - 7)$

$V(7, -2)$; $c = \frac{5}{2}$

20. $(x - 3)^2 = -9(y + 1)$

$V(3, -1)$; $c = \frac{9}{4}$

21. $(x + \frac{1}{3})^2 = -2(y - \frac{4}{3})$

$V(-\frac{1}{3}, \frac{4}{3})$; $c = \frac{1}{2}$

22. $(x + 1)^2 = -\frac{2}{3}(y - \frac{16}{3})$

$V(-1, \frac{16}{3})$; $c = \frac{1}{6}$

23.

24.

25. $y^2 = 10(x - \frac{5}{2})$

26. $x^2 = -6(y - \frac{3}{2})$

27. $(y - 2)^2 = -16(x + 1)$

28. $(x - 4)^2 = 12(y + 1)$

29. $(x + 2)^2 = 24(y + 3)$

30. $(y - 4)^2 = 16(x + 3)$

31. $(x + 3)^2 = -\frac{1}{3}(y - 2)$

32. $(y - 2)^2 = -\frac{36}{7}(x - 4)$

33. $y^2 = -12(x - 3)$

34. $x^2 = 8(y + 2)$

35. $x^2 = 18(y + 9/2)$

36. $y^2 = -4(x - 1)$

37. $y^2 = 8(x + 2)$

38. $y^2 = -6(x - 3/2)$

39. $r = \dfrac{8}{1 + \cos \theta}$ **40.** $r = \dfrac{4}{1 - \cos \theta}$

41. $r = \dfrac{4}{1 - \sin \theta}$ **42.** $r = \dfrac{3}{1 + \cos \theta}$

43. $r \sin^2\theta = 4 \cos \theta$ **44.** $r \cos^2\theta = -2 \sin \theta$

45. $4r^2\cos^2\theta = r \sin \theta - 3$

46. $r \cos \theta + 1 = 2(r \sin \theta - 3)^2$

47. $(0, 0)$ **48.** $(0, 0)$

49. tangent line: $y + 2 = -(x - 1)$ or
$$x + y + 1 = 0$$
normal line: $y + 2 = x - 1$ or
$$x - y - 3 = 0$$

50. $x = 2$ **51.** $3x + y - 5 = 0$

52. $(y + 2)^2 = 6(x - \frac{5}{2})$

53. $x^2 = -6(y + 3)$ **54.** $\frac{2}{3}$ ft or 8 in.

55. $F(\frac{9}{4}, 0)$ **56.** $x^2 = -4cy$

57. $y^2 = 4cx$ **58.** $y^2 = -4cx$

59. $V(0, 0)$; $F(c, 0)$ or $y^2 = 4cx$; $P(c, y_0)$ so
$$y_0{}^2 = 4c^2 \text{ which implies } y_0 = \pm 2c.$$
$$d = \sqrt{(c - c)^2 + (2c - 0)^2} = 2c$$
The focal chord has length $2d$ or $4c$.

60. Let $P(r, \theta)$ be a point on $r = \dfrac{p}{1 - \cos \theta}$. The alternate primary form is $P(-r, \theta + \pi)$. Thus,
$$-r = \frac{p}{1 - \cos(\theta + \pi)}$$
$$r = \frac{-p}{1 - \cos \theta \cos \pi - \sin \theta \sin \pi}$$
$$= \frac{-p}{1 + \cos \theta}$$

61. $y^2 = 4cx$, $x \ge 0$ and $c > 0$. The distance from the vertex to $P(x, y)$ is
$$(x - c)^2 + y^2 = d^2$$
$$(x - c)^2 + 4cx = d^2$$
To minimize d, we minimize d^2, so we take the derivative on the left and set it equal to 0:
$$2x - 2c + 4c = 0$$
$$x = -2c$$
The minimum value occurs when $x = 0$.

62. $y^2 = 4cx$, $x \ge 0$ and $c > 0$. The derivative is $2yy' = 4c$; at $P_1(c, 2c)$ we have
$$m_1 = y' = 2c/y; \text{ at } (c, 2c), m_1 = 1$$
The tangent line is T_1: $y - 2c = x - c$ or $y = x + c$. At $P_2(c, -2c)$,
$$m_2 = y' = 2c/y; \text{ at } (c, -2c), m_2 = -1$$
The tangent line is T_2: $y + 2c = -x + c$ or $y = -x + c$. These tangents intersect when
$$x + c = -x - c$$
$$x = -c$$
If $x = -c$, then $y = 0$ and the ends of the focal chord intersect on the directrix.

63. $A = 4c^2$

64. Let $(x - h)^2 + (y - k)^2 = R^2$ be the equation for the circle and $x^2 = 4cy$ the parabola. The abscissas of the common points satisfy
$$(x - h)^2 + \left(\frac{x^2}{4c} - k\right)^2 = R^2$$
$$x^2 - 2hx + h^2 + \frac{x^4}{16c^2} - \frac{kx^2}{2c} + k^2 = R^2$$
Note the coefficient of x^3 is 0. Suppose the polynomial in the left member has roots x_1, x_2, x_3, and x_4. Then
$$(x - x_1)(x - x_2)(x - x_3)(x - x_4) = 0$$
If we expand the left member the coefficient of x^3 is $-(x_1 + x_2 + x_3 + x_4)$ which is 0 as found above.

65. **a.** $y - y_0 = \dfrac{x_0}{2c}(x - x_0)$

    b. $Q\left(0, y_0 - \dfrac{x_0{}^2}{2c}\right)$

    c. For $|\overline{FP}|^2$ we have
$$x_0{}^2 + (y_0 - c)^2 = \left(\frac{x_0{}^2}{4c} + c\right)^2$$
For $|\overline{FQ}|^2$ we have
$$\left(y_0 - \frac{x_0{}^2}{2c} - c\right)^2 = \left(\frac{x_0{}^2}{4c} + c\right)^2$$
Thus, $|\overline{FP}| = |\overline{FQ}|$ so $\triangle QFP$ is isosceles.

    d. $\phi = \angle FQP = \theta = \angle FPQ$; since L is parallel to $\overline{FQ}$, $\angle LPT = \theta$ also.

Problem Set 5, Page 75

15. $\dfrac{(y-3)^2}{4} - \dfrac{(x-1)^2}{9} = 1$ **16.** $\dfrac{(x+1)^2}{4} - \dfrac{(y-1)^2}{1} = 1$

17. $\dfrac{(x+1)^2}{1/4} + \dfrac{(y-1)^2}{1} = 1$ **18.** $\dfrac{(x+1)^2}{1} + \dfrac{(y-1)^2}{1/4} = 1$

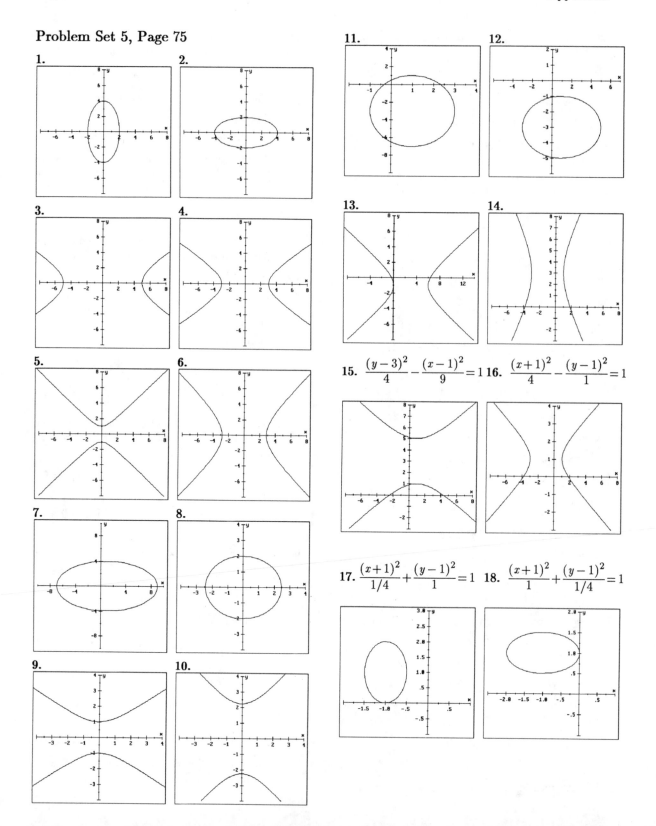

19. $\dfrac{x^2}{4} + \dfrac{(y-1)^2}{9} = 1$　　**20.** $\dfrac{(x-1)^2}{9} + \dfrac{y^2}{4} = 1$

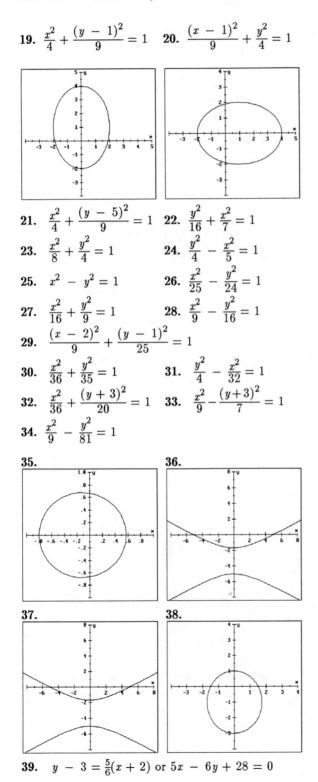

21. $\dfrac{x^2}{4} + \dfrac{(y-5)^2}{9} = 1$　**22.** $\dfrac{y^2}{16} + \dfrac{x^2}{7} = 1$

23. $\dfrac{x^2}{8} + \dfrac{y^2}{4} = 1$　　**24.** $\dfrac{y^2}{4} - \dfrac{x^2}{5} = 1$

25. $x^2 - y^2 = 1$　　　**26.** $\dfrac{x^2}{25} - \dfrac{y^2}{24} = 1$

27. $\dfrac{x^2}{16} + \dfrac{y^2}{9} = 1$　　**28.** $\dfrac{x^2}{9} - \dfrac{y^2}{16} = 1$

29. $\dfrac{(x-2)^2}{9} + \dfrac{(y-1)^2}{25} = 1$

30. $\dfrac{x^2}{36} + \dfrac{y^2}{35} = 1$　　**31.** $\dfrac{y^2}{4} - \dfrac{x^2}{32} = 1$

32. $\dfrac{x^2}{36} + \dfrac{(y+3)^2}{20} = 1$　**33.** $\dfrac{x^2}{9} - \dfrac{(y+3)^2}{7} = 1$

34. $\dfrac{x^2}{9} - \dfrac{y^2}{81} = 1$

35.　　　　　　　　　　**36.**

37.　　　　　　　　　　**38.**

39. $y - 3 = \frac{5}{6}(x+2)$ or $5x - 6y + 28 = 0$

40. There are two cases.
$$\frac{(x-b)^2}{9} + \frac{(y-3)^2}{b^2} = 1$$
$$\frac{(x-a)^2}{a^2} + \frac{(y-3)^2}{9} = 1$$

41. $P(\pm\sqrt{3}, -\frac{1}{2})$　　　**42.** $d = 3 - \sqrt{2}$

43. $\dfrac{(x-1)^2}{4} - \dfrac{4(y+1)^2}{81} = 1$

44. $\dfrac{x^2}{81} - \dfrac{y^2}{81} = 1$　　**45.** $\dfrac{x^2}{4} - \dfrac{y^2}{1} = 1$

46. $\cosh^2 t = \dfrac{x - x_0}{a}$ and $\sinh t = \dfrac{y - y_0}{b}$

Since $\cosh^2 t - \sinh^2 t = 1$, we have
$$\frac{(x-x_0)^2}{a^2} - \frac{(y-y_0)^2}{b^2} = 1$$
provided that $x \geq x_0 + a$ because $\cosh t \geq 1$.

47. $\sinh^2 t = \dfrac{x - x_0}{a}$ and $\cosh t = \dfrac{y - y_0}{b}$

Since $\cosh^2 t - \sinh^2 t = 1$, we have
$$\frac{(y-y_0)^2}{b^2} - \frac{(x-x_0)^2}{a^2} = 1$$
provided that $y \geq y_0 + b$ because $\cosh t \geq 1$.

48. $y = \pm x$ are the asymptotes, which are perpendicular to each other (their slopes are negative reciprocals). If the asymptotes are perpendicular to each other, their slopes are negative reciprocals and $m_1 = -1/m_2 = m$. Since the asymptotes make an angle of $\pm\pi/4$ with the positive x-axis, $m = 1$. The equations of the asymptotes then are $y = x + b$ and $y = -x + b$, with $b = 0$ because of standard position. Then $(y + x)(y - x) = K$ is the equation of the hyperbola $y^2 - x^2 = K$. The point $(a, 0)$ is a vertex, so $K = -a^2$ and $x^2 - y^2 = a^2$.

49. $\dfrac{x^2}{50^2} + \dfrac{y^2}{40.5^2} = 1$; The foci are $2c \approx 58.643$ million miles apart.

50. 9.1419×10^7 mi and 9.4581×10^7 mi.

51. Let d_1 be the distance that sound travels from the gun at A to the person at P. Let d_2 be the distance that sound travels from the gong to B to the person at P. Let x be the distance that it takes the bullet to reach the gong at B. Then, $d_1 - d_2 = x$ and the person should stand on a branch of the hyperbola with foci at A and B.

52. The airplane is located at point $(3.5387, \frac{\pi}{8})$.

53. Let d_1 be the distance from $P(x, y)$ to $F_1(-c, 0)$ and d_2 be the distance from $P(x, y)$ to $F_2(c, 0)$. By definition,

$$d_1 + d_2 = 2a$$
$$d_1 = 2a - d_2$$
$$d_1^2 = 4a^2 - 4ad_2 + d_2^2$$
$$(x + c)^2 + y^2 = 4a^2 - 4ad_2 + 4(x - c)^2 + y^2$$
$$2cx = 4a^2 - 4ad_2 - 2cx$$
$$cx = a^2 - ad_2$$
$$ad_2 = a_2 - cx$$
$$a^2[(x - c)^2 + y^2] = a^4 - 2a^2cx + c^2x^2$$
$$a^2x^2 - 2a^2c + a^2c^2 + a^2y^2 = a^4 - 2a^2cx + c^2x^2$$
$$(a^2 - c^2)x^2 + a^2y^2 = a^2(a^2 - c^2)$$
$$b^2x^2 + a^2y^2 = a^2b^2$$
$$\frac{x^2}{a^2} + \frac{y^2}{b^2} = 1$$

54. Let d_1 be the distance from $P(x, y)$ to $F_1(-c, 0)$ and d_2 be the distance from $P(x, y)$ to $F_2(c, 0)$. By definition,

$$d_1 - d_2 = 2a$$
$$d_1 = 2a + d_2$$
$$d_1^2 = 4a^2 + 4ad_2 + d_2^2$$
$$(x + c)^2 + y^2 = (x - c)^2 + y^2 + 4a^2 + 4ad_2$$
$$2cx = 4a^2 + 4ad_2 - 2cx$$
$$cx = a^2 + ad_2$$
$$ad_2 = -a_2 + cx$$
$$a^2[(x - c)^2 + y^2] = a^4 - 2a^2cx + c^2x^2$$
$$a^2x^2 - 2a^2cx + a^2c^2 = a^4 - 2a^2cx + c^2x^2$$
$$(a^2 - c^2)x^2 + a^2y^2 = a^2(a^2 - c^2)$$
$$-b^2x^2 + a^2y^2 = -a^2b^2$$
$$\frac{x^2}{a^2} - \frac{y^2}{b^2} = 1$$

55. **a.** $A = C = 0$
   b. A and C have the same signs
   c. $A = C$
   d. $A = C$ and $D = E = 0$
   e. A and C have opposite signs
   f. $A > 0$ and

$$\frac{D^2}{4A} + \frac{E^2}{4C} < F$$

or

$A < 0$ and

$$\frac{D^2}{4A} + \frac{E^2}{4C} > F$$

56. **a.**
$$\frac{x^2}{a^2} + \frac{y^2}{b^2} = 1$$
$$\frac{2x}{a^2} + \frac{2yy'}{b^2} = 0$$
$$y' = -\frac{b^2x}{a^2y}$$

The slope at $P_0(x_0, y_0)$ is

$$m = y' = -\frac{b^2x_0}{a^2y_0}$$

The equation of the tangent line is

$$y - y_0 = -\frac{b^2x_0}{a^2y_0}(x - x_0)$$
$$\frac{y_0y}{b^2} - \frac{y_0^2}{b^2} + \frac{x_0x}{a^2} - \frac{x_0^2}{a^2} = 0$$
$$\frac{x_0x}{a^2} + \frac{y_0y}{b^2} = \frac{x_0^2}{a^2} + \frac{y_0^2}{b^2}$$
$$\frac{x_0x}{a^2} + \frac{y_0y}{b^2} = 1$$

(Since $P_0(x_0, y_0)$ lies on the ellipse.)

   b. At $P_0(\pm a, 0)$,
$$\frac{\pm ax}{a^2} = 1$$
or $x = \pm a$, a vertical tangent line.

At $P_0(0, \pm b)$
$$\frac{\pm by}{b^2} = 1$$
or $y = b$, a horizontal tangent line.

57. From the hyperbola
$$y = \pm\frac{b}{a}\sqrt{x^2 - a^2}$$

and from the asymptote $y = \pm\frac{b}{a}x$. Let d be the vertical distance between points on the hyperbola, and on the asymptote x. Then,

$$\lim_{x \to \infty} \frac{b}{a}(\sqrt{x^2 - a^2} - x)$$
$$= \lim_{x \to \infty} \frac{b}{a}\left[\frac{(\sqrt{x^2 - a^2} - x)(\sqrt{x^2 - a^2} + x)}{\sqrt{x^2 - a^2} + x}x\right]$$

$$= \lim_{x \to \infty} \frac{b}{a}\left[\frac{x^2 - a^2 - x^2}{\sqrt{x^2 - a^2} + x}\right]$$

$$= -ab \lim_{x \to \infty} \frac{1}{\sqrt{x^2 - a^2} + x} = 0$$

Problem Set 6, Page 92

1.   2.

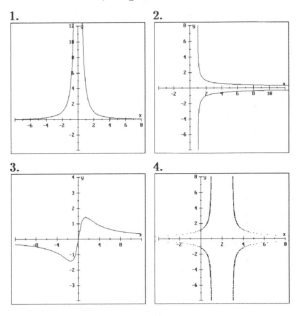

3.   4.

5. Symmetric with respect to the origin.
    Extent: domain, all $x \neq 0$; range, all $y \neq 0$.
    Asymptotes: $x = 0$, $y = 0$.
    No intercepts.
6. Symmetric with respect to the origin.
    Extent: domain, all $x \neq 0$; range, all $y \neq 0$.
    Asymptotes: $x = 0$, $y = 0$.
    No intercepts.
7. No symmetry with respect to the x-axis,
    y-axis, or origin.
    Extent: domain, all $x \neq 0$; range, all $y \neq 1$.
    Asymptotes: $x = 0$, $y = 1$.
    Intercept: $(-1, 0)$.
8. No symmetry with respect to the x-axis,
    y-axis, or origin.
    Extent: domain, all $x \neq -2$;
        range, $y \neq -1$.
    Asymptotes: $x = -2$, $y = 1$
    Intercepts: $(0, \frac{1}{2})$, $(-1, 0)$.
9. No symmetry with respect to the
    x-axis, y-axis, or origin.

Extent: domain, all $x \neq -2$; range, all real
    numbers.
Asymptotes: $x = -2$, $y = 2x - 3$.
Intercepts: $(0, -5)$, $(-\frac{5}{2}, 0)$, $(2, 0)$.

10. No symmetry with respect to the
    x-axis, y-axis, or origin.
    Extent: domain, all $x \neq -2$; range, all
        $y \neq -7$.
    Intercepts: $(0, -1)$, $(\frac{1}{3}, 0)$.
11. No symmetry with respect to the x-axis,
        y-axis, or origin.
    Extent: domain, all $x \neq -\frac{1}{2}$; range,
        $y \geq -1$, $y \neq \frac{5}{4}$.
    Intercepts: $(0, 0)$, $(2, 0)$.
12. No symmetry with respect to the x-axis,
        y-axis, or origin.
    Extent: domain, all $x \neq -2$; range, all
        $y \neq -3$.
    Intercept: $(0, 7)$.
13. Symmetric with respect to the x-axis, y-axis,
        and origin.
    Extent: domain, $-2 \leq x \leq 2$; range
        $-3 \leq y \leq 3$;
    Intercepts: $(-2, 0)$, $(2, 0)$, $(0, 3)$, $(0, -3)$.
14. Symmetric with respect to the x-axis, y-axis,
        and origin.
    Extent: domain, all real numbers; range,

    $$y \geq \sqrt{5} \text{ or } y \leq -\sqrt{5}.$$

    Asymptotes: $y = \pm\sqrt{3}\,x$

    Intercepts: $(0, \sqrt{5})$, $(0, -\sqrt{5})$.
15. Symmetric with respect to the origin. Extent:

    $$\text{domain, } -\frac{\sqrt{26}}{2} \leq x \leq \frac{\sqrt{26}}{2}; \text{ range,}$$

    $$-\frac{\sqrt{26}}{2} \leq y \leq \frac{\sqrt{26}}{2}.$$

    Intercepts: $(0, \pm\frac{6}{13}\sqrt{26})$, $(\pm\frac{6}{13}\sqrt{26}, 0)$.
16. Symmetric with respect to the x-axis.
    Extent: domain, $x < 2$; range, all reals,
        $y \neq 0$.
    Asymptotes: $x = 2$, $y = 0$.
    Intercepts: $(0, 1)$, $(0, -1)$.
17. No symmetry with respect to the x-axis, y-
        axis, or origin.
    Extent: domain, all reals $x \neq 3$, $x \neq 1$;
        range, $y < -4$ or $y > 0$.
    Asymptotes: $x = 3$, $x = 1$, $y = 0$.
    Intercept: $(0, \frac{4}{3})$.

18. No symmetry with respect to the x-axis, y-axis, or origin.

Extent: domain, $x \geq -2^{2/3}$; range, all real numbers.

Asymptotes: no horizontal or vertical asymptotes.

Intercepts: $(0, 0)$, $(0, -4)$.

19. Symmetric with respect to the x-axis; y-axis, and origin.

Extent: domain, $x \leq -2$ or $-1 < x < 1$ or $x \geq 2$; range, all real numbers.

Asymptotes: $x = 1$, $x = -1$.

Intercepts: $(0, 0)$, $(2, 0)$, $(-2, 0)$.

20. Symmetric with respect to the x-axis.

Extent: domain, $x \leq 1$ or $x > 2$; range, all reals, $y \neq 1$, $y \neq -1$.

Asymptotes: $x = 2$, $y = 1$, $y = -1$.

Intercepts: $(0, \frac{1}{2}\sqrt{2})$, $(0, -\frac{1}{2}\sqrt{2})$, $(1, 0)$.

21. rotated hyperbola; $\theta = 45°$

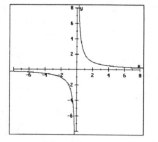

22. rotated hyperbola; $\theta = 45°$

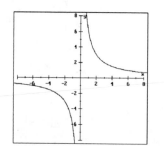

23. rational function $y = 1/x$, which has been translated to the point $(0, 1)$

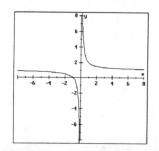

24. rational function $y = 1/x$, which has been translated to the point $(-2, 1)$

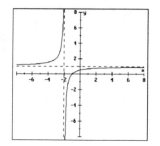

25. rational function

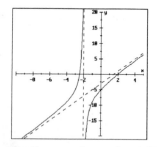

26. the line $y = 3x - 1$, with a deleted point at $x = -2$

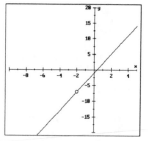

27. a parabola $y = x^2 - 2x$, with a deleted point at $x = -\frac{1}{2}$

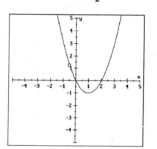

28. a parabola $y = x^2 + 4x + 7$, with a deleted point at $x = -2$

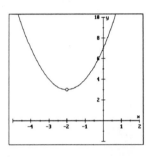

29. Recognize as an ellipse: $\dfrac{x^2}{4} + \dfrac{y^2}{9} = 1$.

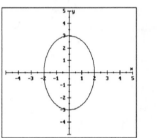

30. Recognize as a hyperbola: $\dfrac{y^2}{5} - \dfrac{x^2}{\frac{5}{3}} = 1$.

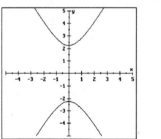

31. vertical ellipse with 45° rotation; if you carry through the rotation, the equation is

$$\frac{x'^2}{9} + \frac{y'^2}{4} = 1.$$

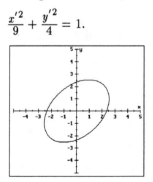

32. See answer to Problem 16 for a description.

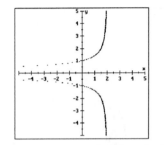

33. See answer to Problem 17 for a description.

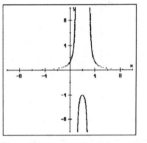

34. See answer to Problem 18 for a description.

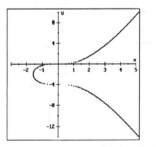

35. See answer to Problem 19 for a description.

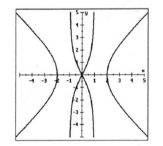

36. See answer to Problem 20 for a description.

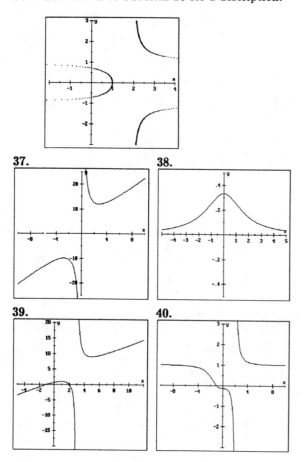

37.

38.

39.

40.

C.

Index